P W Williams
University of Manchester Institute of Science
and Technology

Numerical computation

Nelson

Thomas Nelson and Sons Ltd
36 Park Street London W1Y 4DE

Nelson (Africa) Ltd
PO Box 18123 Nairobi Kenya

Thomas Nelson (Australia) Ltd
171–175 Bank Street South Melbourne Victoria 3205

Thomas Nelson and Sons (Canada) Ltd
81 Curlew Drive Don Mills Ontario

Thomas Nelson (Nigeria) Ltd
PO Box 336 Apapa Lagos

First published in Great Britain by Thomas Nelson and Sons Ltd, 1972
Reprinted with amendments 1973

ISBN 0 17 771018 7

Printed in Great Britain by
The Camelot Press Ltd, London and Southampton

John Senior

Numerical computation

Contents

Preface *vii*

1 Computer calculations

1.1 Introduction *1* **1.2** Binary arithmetic *3* **1.3** Representation of numbers in a computer *5* **1.4** Errors in arithmetic operations *6* **1.5** Errors of computational methods *9* **1.6** Acceleration of convergence *12* Bibliographic notes *14* Worked examples *16* Problems *22*

2 Solution of nonlinear equations

2.1 A simple iteration example *24* **2.2** Finding initial approximations *25* **2.3** Simple iteration *26* **2.4** Bisection method *27* **2.5** Newton's method *28* **2.6** Secant method *31* **2.7** Method of false position *32* **2.8** Comparison of methods *32* **2.9** Systems of nonlinear equations *34* Bibliographic notes *35* Worked examples *35* Problems *37*

3 Solution of polynomial equations

3.1 Introduction *38* **3.2** Polynomial arithmetic *39* **3.3** Finding initial approximations *41* **3.4** Complete solution of a polynomial *43* **3.5** Methods for finding all roots *46* **3.6** Comparison of methods *51* Bibliographic notes *51* Worked examples *51* Problems *56*

4 Solution of linear simultaneous equations

4.1 Introduction *57* **4.2** Direct methods *58* **4.3** Iterative methods *67* **4.4** Sparse matrices *70* **4.5** Comparison of methods *71* Bibliographic notes *72* Worked examples *73* Problems *78*

5 Solution of ordinary differential equations

5.1 Introduction *80* **5.2** Initial-value problem *81* **5.3** Predictor–corrector methods *84* **5.4** Runge–Kutta methods *91* **5.5** Stability *94* **5.6** Further methods *98* **5.7** Comparison of methods *100* Bibliographic notes *102* Worked examples *102* Problems *105*

6 Finite differences

6.1 Behaviour of finite differences *107* **6.2** Errors in finite-difference tables *109* **6.3** Finite-difference operators *110* **6.4** Interpolation *112* **6.5** Formulae for differentiation and integration *114* **6.6** Solution of linear difference equations *115* Bibliographic notes *116* Worked examples *116* Problems *118*

7 Curve fitting

7.1 Interpolation and approximation *119* **7.2** Interpolation *120* **7.3** Fitting by the method of least squares *125* **7.4** Orthogonal polynomials *128* **7.5** Chebyshev polynomials and minimax approximation *132* Bibliographic notes *138* Worked examples *139* Problems *144*

8 Numerical integration

8.1 Introduction *145* **8.2** Newton–Cotes formulae *146* **8.3** Gaussian quadrature *151* **8.4** Romberg integration *154* **8.5** Comparison of methods *155* Bibliographic notes *156* Worked examples *156* Problems *159*

9 Eigenvalues and eigenvectors

9.1 Fundamental equations *160* **9.2** Some useful matrix results *161* **9.3** Iterative methods *162* **9.4** Transformation methods *167* **9.5** Eigenvalues of a tridiagonal matrix *170* **9.6** Further methods *171* Bibliographic notes *172* Worked examples *173* Problems *182*

Bibliography 183

Answers to problems 185

Index 189

Preface

The importance of numerical computation to the scientist and engineer is now widely acknowledged and most undergraduates have some training in the use of computers and have access to a computer for the solution of their problems. There is growing awareness that a knowledge of the various computational methods which are available, and an appreciation of their strong points and pitfalls, is essential. Of the various texts which are available, some concentrate on giving the definitive body of mathematical knowledge associated with particular methods and are too advanced for the undergraduate engineer. Others concentrate on the learning of a particular programming language and often cover only a very limited range of methods. The purpose of this text is to introduce a wide range of numerical methods at a level suitable to the non-specialist.

The text has the following aims:

1. To describe the various methods available for the problems considered. In the case of the simpler methods this description is full enough for the student to flow-chart and program the method himself. In the case of the more complex methods, a reference to the detailed description in other literature is given, together with a more general description of the motivation and usefulness of the method.
2. To develop an appreciation of the advantages and disadvantages of the various methods and the type of problem to which they are suited.
3. To give arithmetic examples of the easier methods, and further problems on these for the student to develop his skill in numerical computation.
4. To indicate how computer errors arise and propagate and show how these considerations affect the accuracy of the methods discussed.

The text is not suitable as a mathematical textbook; it is intended more as a guide book to the practice of numerical computation. Mathematical theorems are quoted without proof in those cases where the result is important but the proof obscure. Proofs are included only where they illuminate some important principle or property.

Computer programs are not included in the text but the understanding of the methods and an appreciation of the problems of numerical computation will be enhanced by a knowledge of programming. It is therefore desirable, but not essential, if readers of the book are familiar with computer programming.

The text presupposes a knowledge of mathematics such as that normally given on an engineering course in higher education, i.e., a knowledge of calculus up to Taylor series and the mean-value theorem. For some sections a knowledge of matrix and vector algebra is necessary. References to appropriate text books for this background knowledge are given where relevant. The chapter on eigenvalues is the most difficult from the point of view of the background material and some readers will find it necessary either to take some theory on trust or ignore this chapter.

The material is suitable for undergraduate students, and others of a similar level and, in the climate of increasing demand for numerical computation, such material may well become essential for all engineering and science students. Much of the material is better placed in the second and third year of study but the simpler topics could be commenced in the first year. At the present time this text is also suited to postgraduate work since numerical computation is not yet included for many undergraduates.

It is envisaged that many students will move on to specialize in particular topics and, for this purpose, bibliographic notes are included at the end of each chapter. This will enable the reader either to pursue further reading at an elementary level or to move on to more specialized texts as his interest and competence increase. References in the text are by author with the year of publication and bibliographic details tabled in alphabetical order at the end of the book.

In order to maintain continuity of the text many of the worked examples are grouped at the end of each chapter together with problems for the student to attempt. The examples contain valuable additional material and it is felt that the reader will not gain full benefit from the text unless the examples are also studied. As well as illustrating computational procedure several of the examples show situations in which a method may be unsuitable.

I should like to acknowledge the assistance of my colleagues Gerard Conroy and Dennis Clough in reading the manuscript and making several corrections and helpful suggestions. Thanks are also due to the latter for his help in the preparation of some of the problems, examples, and the computer programming which this involved. Finally, my sincere thanks to the typists, Mrs Betty Sharples and Miss Jane Warden for the preparation of the typescript. I should also like to thank Professor L. Fox for his helpful corrections and suggestions for improvements in the text.

P W W

1

Computer calculations

1.1 Introduction

Before embarking on a study of computational methods for the computer, it is perhaps worthwhile to consider why the use of computers for scientific calculations has increased so dramatically over the last 20 years. In the past, the human brain has been shown capable of impressive mathematical developments, so where does the advantage of the computer lie? Although the brain is capable of inter-relating a vast range of thought and retained knowledge, when it comes to performing a simple calculation, such as $19 \cdot 3281 \times 2654 \cdot 3$, our mental apparatus is both slow and imperfect. The average reader would probably take about 2 minutes to perform the above calculation, and the answer he produced would very likely be wrong! In the same period of time the most recent computers could perform 500 000 000 such calculations, and only a breakdown of the electronic equipment would produce the equivalent of a human mistake. It is perhaps unfortunate that the answer produced is nearly always inaccurate, but since this inaccuracy is limited to something like one part in 10^{12}, one might think that this is acceptable. The importance of this inevitable inaccuracy is one of the topics which will constantly recur in the rest of the text.

One of the advantages of the computer then is speed and reliability in the execution of simple operations. The other major advantage lies in the stimulus which has been given to new computational methods. Although the body of mathematical knowledge is constantly developing, there are many problems which cannot be solved by mathematical analysis. For example, if one considers the topic of integration there are many functions which have no formula for their integral. The only solution to such problems is to use numerical approximation, which uses precisely the kind of repetitive arithmetic for which a computer is ideal. Indeed, the speed of modern computers has made possible the use of numerical approximation techniques in physical problems which would otherwise be quite impossible to solve owing to the size of the numerical computation problem which results. This type of problem arises, for example, when a continuous substance, such as a body of fluid or a solid beam, is broken up into small units in order to make the mathematical analysis feasible. Some of the problems now being computed take several hours on the fastest modern machines and the demand is for still better approximation which requires vastly increased computing time.

It will be advantageous to obtain some ideas on how calculations are carried out on a computer, so that the problems and limitations of numerical computation, which are discussed in the ensuing chapters, can be more easily appreciated. It is easy to see why the modern computer is based on electricity. The speed of transmission of electricity, while not being instantaneous, is certainly near the limit of what is physically possible. The on–off two-state property of electrical signals is sufficient to represent numbers in the binary-number system, and the development of transistors has brought switching, storing, and timing devices, all of which operate at high speeds. Indeed, the speed of these components is now such that the time taken for the electricity to pass down the wires is a serious cause of delay, and recent design activity has centred on eliminating wiring by miniaturizing components, and manufacturing these in groups already linked together.

The study of the electronic circuitry which makes the modern computer possible is clearly beyond the scope of this book, but we will consider here some of the stages in the process in order to give an insight into how the calculations proceed. First, the method of calculation must be precisely defined and a sequence of instructions prepared which can be fed into the computer. In the early days of computers this was very tedious because the mathematical steps had to be broken down, or translated, into the simple logical steps which the computer can perform. Programming a computer was by machine-code instructions which is the name for the fundamental instructions which a computer can obey. Clearly, from the point of view of the scientist or engineer, this extra translation process was a barrier to effective use of the machines and so the translation process was switched to the machine itself. High-level languages, such as ALGOL or FORTRAN were introduced round about 1960. These were designed to be close to the scientist's normal way of specifying a method, and yet each instruction could be translated into a set of basic machine-code instructions. Thus, when a high-level language program is read into a computer a translation program, called a compiler, breaks down the high-level instructions into the machine code.

Certain problems also exist when considering number representation within a computer. The familiar decimal notation, involving 10 different digits, is not the most suitable form for this representation, since electronic devices would be required which could generate, and recognize, 10 different states. While this would be possible, the complexity of the devices would make them prohibitively expensive and susceptible to error. However, it is possible to build very simple electronic devices which respond to a two-state signal indicating on or off. With these devices it is possible to store a sequence of bits of binary information and combine these in various ways to represent operations such as addition, subtraction, etc. Fortunately, a number system exists which requires only two basic symbols, 0 and 1, in which numbers are represented as a string of zeros and ones. This number system, the binary system, is therefore used for the representation of numbers inside a computer. This does mean that a translation process is necessary when numbers are read into, or output from, a computer, but this process is relatively trivial. The relations between the binary and decimal numbers are discussed in Section 1.2.

There is a further problem encountered when attempting to store numbers in a computer which is a feature of both the decimal and binary-number systems. If one considers numbers in the decimal system such as $\frac{1}{3}$ or *e*, it is not possible to represent these using a finite number of decimal places. In a similar fashion there are many numbers which cannot be represented by a finite number of binary digits. Since a computer only has a limited number of binary digits to represent a single number (typically between 24 and 60) then, any number which requires more digits than this will inevitably be represented with some loss of accuracy.

Now, we come to the calculations which are possible in simple machine instructions. Addition and subtraction can be arranged as simple combinations of electrical signals and present no problems. However, if one considers the multiplication of two large numbers, a little analysis is necessary to see how we ourselves perform such a calculation. Essentially, there are three operations involved; namely, multiplication by a single digit using remembered tables, changing columns to allow for units, tens, hundreds, etc., and the subsequent addition of the various sub-totals. The first of these is trivial in binary since there are only three entries in the multiplication table. The second is a standard machine-code operation so that the process becomes a series of place shifts and additions.

The use of multiplication is so frequent that a special piece of electronic equipment is often used to do this, rather than creating a set of machine-code instructions on every occasion. Similar arguments apply to division which becomes a series of shifts and subtractions.

Any other operations such as taking logarithms, powers, or forming trigonometric functions must in some way be broken down into the basic operations as described above. A set of machine-code instructions is then used to execute such functions. The method of calculation for these functions involves approximation methods so that further errors are introduced by the use of functions such as the above. To complete the picture, there are control units in the computer which supervise the input and output of information to the computer and the movements between the store and the calculating unit. The function of one of these control units is to use the instructions stored in the computer to actuate the machine-code instructions which effect the desired calculations. It is clear from the above discussion that familiarity with the binary number system and computer errors is essential background for numerical computation and these topics will be investigated in the remainder of this chapter.

1.2 Binary arithmetic

1.2.1 The binary system

The best introduction to the binary system, which is the basis of computer operation, is to look carefully at the more familiar decimal system. This is based on columns representing the powers of 10 so that 2304 means

$$2 \times 10^3 + 3 \times 10^2 + 0 \times 10^1 + 4 \times 10^0$$

or in general terms, a decimal number

$$a_n a_{n-1} \ldots a_1 a_0 \qquad 0 \leqslant a_i \leqslant 9 \tag{1.1a}$$

represents the decimal number

$$a_n \times 10^n + a_{n-1} \times 10^{n-1} + \cdots + a_1 \times 10 + a_0 \qquad (1.1b)$$

It is possible, of course, to have a base other than 10. Thus, if we have a number system to the base r, a number

$$a_n a_{n-1} \ldots a_1 a_0 \qquad 0 \leqslant a_i \leqslant r - 1 \qquad (1.2a)$$

represents the decimal number

$$a_n \times r^n + a_{n-1} \times r^{n-1} + \cdots + a_1 \times r + a_0 \qquad (1.2b)$$

Choosing the base 2, we see that 10110 represents in the decimal system

$$1 \times 2^4 + 0 \times 2^3 + 1 \times 2^2 + 1 \times 2^1 + 0 \times 2^0 = 22$$

The most striking difference between the two systems is the number of digits required to write down a number. In the decimal system we use the digits $0, \ldots, 9$ and any figure higher than 9 is broken down into units of the base 10. In the binary system, the digits less than the base 2 are 0 and 1 and any higher number automatically involves multiples of 2 in the higher columns. Thus, the first few numbers are as follows:

$$
\begin{array}{ll}
0_{10} = 0_2 & 3_{10} = 11_2 \\
1_{10} = 1_2 & 4_{10} = 100_2 \\
2_{10} = 10_2 & 5_{10} = 101_2 \quad \text{etc.}
\end{array}
$$

The subscript notation will be used to indicate the base used when there is any possibility of confusion. The reader who is unfamiliar with the binary system should try some of the simple examples at the end of this chapter. (See Examples 1.1, 1.2, 1.3, and 1.4.) Details of arithmetic operations in binary can be found in several texts, for example, Conte (1965) but attention here will be restricted to the conversion of numbers from binary to decimal and vice versa.

1.2.2 Binary to decimal conversion

As an example, take the binary number 101100. One way in which this could be converted is to give each column its equivalent decimal value and sum the parts, i.e., $32 + 8 + 4 = 44$. There is, however, a simpler scheme which is very repetitive and hence suitable for computer use. Let the digits be a_r $(r = 0, 1, \ldots)$ numbering from the *right* to the *left*. Starting from the left-most digit a_k, compute as follows:

$$
\begin{array}{ll}
b_k = a_k & (1.3a) \\
b_r = 2b_{r+1} + a_r, \qquad r = k - 1, \qquad k - 2, \ldots, 0 & (1.3b)
\end{array}
$$

In the above example this gives

$$
\begin{array}{l}
b_5 = 1 \\
b_4 = 1 \cdot 2 + 0 = 2 \\
b_3 = 2 \cdot 2 + 1 = 5 \\
b_2 = 2 \cdot 5 + 1 = 11 \\
b_1 = 2 \cdot 11 + 0 = 22 \\
b_0 = 2 \cdot 22 + 0 = 44
\end{array} \qquad (1.4)
$$

(See also Example 1.1.)

1.2.3   Decimal to binary conversion

The direct approach here is to subtract the largest possible exact power of 2 from the decimal number, make an entry in the appropriate column, and repeat this process until the number remaining is zero. For example, 100_{10} has 64, 32, and 4 subtracted, so that the appropriate binary number is 1100100.

The alternative scheme takes a decimal number b_0 and divides by 2 storing the remainders as a_0, a_1, etc.

$$b_r = \frac{b_{r-1} - a_{r-1}}{2} \tag{1.5}$$

where a_{r-1} is 1 when b_{r-1} is odd and a_{r-1} is 0 when b_{r-1} is even. In the example above we have

$$
\begin{array}{cl}
b_r & a_r \\
2)\ 44 & \text{rem} . 0 \\
2)\ 22 & \text{rem} . 0 \\
2)\ 11 & \text{rem} . 1 \\
2)\ 5 & \text{rem} . 1 \\
2)\ 2 & \text{rem} . 0 \\
2)\ 1 & \text{rem} . 1 \\
0 &
\end{array}
\tag{1.6}
$$

Note, that the left-most binary digit occurs at the bottom of the column of remainders so that the binary number is 101100. (See also Example 1.2.)

1.3   Representation of numbers in a computer

There are two basic types of number representation in a computer, integers and real numbers, with differing arithmetic operations corresponding to these. The integer operations are simpler and quicker, so integer arithmetic should be used where possible. However, in most scientific calculations not only will fractions be required but the calculations will range from very small numbers to very large numbers. In this case, floating-point arithmetic is used which can deal with both these problems.

Floating-point representation of a number in binary form is very similar to the exponent form of a decimal number, e.g., $0 \cdot 2342 \times 10^{-3}$ which has a mantissa of $0 \cdot 2342$ and an exponent of -3. A floating-point machine number has the form

$$\bar{x} = a \times 2^b \tag{1.7}$$

where

$$a = \sum_{r=1}^{t} d_r 2^{-r} \quad \text{and} \quad \tfrac{1}{2} \leqslant |a| < 1$$

and usually $|b| \leqslant M$.

The d_r are the binary digits, 0 or 1, associated with each power of 2.

The values of t and M vary but typical values would be $t = 37$ and $M = 2^8 = 256$. This gives approximately 11 significant decimal digits and a range of numbers between 10^{-77} and 10^{+77}.

The bar notation on the x is used, when necessary, to emphasize the fact that a machine number may not exactly represent the true value. The number may be in error by, at most, one unit in the first column not represented in the mantissa a, i.e., 2^{-t-1}. So the possible error in $\bar{x}$, due to machine round-off, is 2^{b-t-1} to take account of the exponent b.

In general, integer arithmetic is reserved for operations such as counting, since if integers are successively multiplied the number of binary places required to represent the product will grow indefinitely. In the case of floating-point arithmetic, if a large number is produced the exponent is adjusted to represent the correct size and, if the number of digits is greater than t, then the surplus figures are discarded at the least significant end. The limit of the system now occurs when the size of the exponent is exceeded, a condition known as overflow. If surplus digits are discarded, further round-off error is introduced.

Thus we see that errors can be introduced in the computer both by the representation in floating-point form and also by the subsequent calculations. Since floating-point arithmetic is virtually essential for scientific calculations it is useful to have some means of minimizing these errors. The obvious way of doing this is to increase the number of significant figures which means increasing the size of t. Such a scheme is not employed in all calculations because this takes up extra computer time and, for some computers, special programs must be written to provide this facility. However, it is fairly common for double-length working to be provided in modern computing systems. In this case, the errors introduced would be no larger than 2^{b-2t-1} with an obvious increase in accuracy. Many scientific problems lead to calculations involving complex numbers and there is no direct representation of these in a computer. Complex number facilities are provided in some high-level languages such as FORTRAN but users should realize that this is done by means of specially written pieces of program which makes these facilities time-consuming.

1.4   Errors in arithmetic operations

The discussion in this section will be restricted to floating-point arithmetic and will be mainly concerned with the errors involved in the four basic rules of arithmetic. We must consider the errors caused by the finite representation within the machine, known as rounding errors, and also the way in which the various arithmetic operations propagate these errors.

First, we must remember that representing a number x in the computer gives a number $\bar{x}$ with a maximum possible error of $\pm 2^{b-t-1}$. We now consider the addition of two machine numbers:

$$\bar{x}_1 = a_1 . 2^{b_1} \qquad \bar{x}_2 = a_2 . 2^{b_2}, \qquad b_1 > b_2 \tag{1.8}$$

where the machine answer is $\bar{x}_3 = a_3 . 2^{b_3}$. The addition is performed by first shifting the digits of a_2 to the right by $b_1 - b_2$ places so that corresponding columns line up. This is normally done by using working registers which are double length so that no digits are lost in the shift process. The two numbers are then added together. If the resultant sum has a value of a_3 outside the range $\frac{1}{2} \leqslant |a_3| < 1$, then a_3 is moved to the right or left the appropriate number of places and the size of the exponent b_3 is adjusted accordingly. The number is then reduced to single length by discarding the least significant digits. The maximum error introduced by the rounding process is therefore 2^{b_3-t-1}.

In our error calculations it will more often be convenient to consider the relative error which we will define as the absolute error divided by the computed result. A more fundamental definition of relative error is obtained by dividing by the true answer. However, the computed result is the only figure which is available and so the slightly different definition of relative error used here is more useful. Thus, in this case the relative error is $2^{b_3-t-1}/(a_3 . 2^{b_3})$. Since the lowest possible value of a_3 is $\frac{1}{2}$, the maximum value of this quantity is $\pm 2^{-t}$. Similar arguments apply to the execution of the other three processes so that the machine answer to a calculation involving exact machine numbers is

$$\bar{A} = A(1 + e) \tag{1.9a}$$

where

$$|e| \leqslant 2^{-t} \tag{1.9b}$$

and A is the correct answer which cannot normally be represented exactly in single-length working. The notation ε will be used for absolute error and e for relative error.

In general, the calculations do not involve exact machine numbers owing to the effect of previous errors of representation or calculation. It is, therefore, necessary to consider how these errors propagate in the various arithmetic processes. Let

$$x = \bar{x} + \varepsilon_x \qquad y = \bar{y} + \varepsilon_y \tag{1.10}$$

Then

$$x + y = \bar{x} + \bar{y} + \varepsilon_x + \varepsilon_y \tag{1.11}$$

so that the new error is the sum of the individual errors and the maximum value of the modulus is $|\varepsilon_x| + |\varepsilon_y|$. The relative error has the value

$$\text{Relative error} = \frac{\bar{x}}{\bar{x} + \bar{y}} \cdot \frac{\varepsilon_x}{\bar{x}} + \frac{\bar{y}}{\bar{x} + \bar{y}} \cdot \frac{\varepsilon_y}{\bar{y}} \tag{1.12}$$

which is a sum of easily calculated multiples of the relative errors in the original numbers.

The derivation for subtraction is very similar but with negative signs on the right-hand side. It is important to note that in a subtraction sum when $\bar{x}$ and $\bar{y}$ are close, the two multipliers of the relative errors can be very large and this operation can easily produce substantial errors. (See Example 1.5.)

For multiplication

$$xy = (\bar{x} + \varepsilon_x)(\bar{y} + \varepsilon_y)$$
$$= \bar{x}\bar{y} + \varepsilon_x . \bar{y} + \varepsilon_y . \bar{x} + \varepsilon_x . \varepsilon_y \tag{1.13}$$

Normally, the errors ε_x and ε_y will be small so that the product $\varepsilon_x . \varepsilon_y$ can be neglected. The absolute error is then

$$\varepsilon_x \bar{y} + \varepsilon_y \bar{x} \tag{1.14}$$

and the relative error is

$$\frac{\varepsilon_x}{\bar{x}} + \frac{\varepsilon_y}{\bar{y}} \tag{1.15}$$

The relative error in a multiplication approximately equals the sum of the relative errors in the original two numbers.

We will find that a similar result holds for division but in this case we have the difference of the original relative errors. The proof for the division result is obtained by expanding the quotient using the binomial series and neglecting terms involving the product of errors

$$\frac{x}{y} = \frac{\bar{x} + \varepsilon_x}{\bar{y} + \varepsilon_y}$$

$$= \frac{\bar{x}}{\bar{y}}\left(1 + \frac{\varepsilon_x}{\bar{x}}\right)\left(1 + \frac{\varepsilon_y}{\bar{y}}\right)^{-1}$$

$$= \frac{\bar{x}}{\bar{y}}\left(1 + \frac{\varepsilon_x}{\bar{x}}\right)\left(1 - \frac{\varepsilon_y}{\bar{y}} + \left(\frac{\varepsilon_y}{\bar{y}}\right)^2 \cdots\right)$$

$$\approx \frac{\bar{x}}{\bar{y}} + \frac{\varepsilon_x}{\bar{y}} - \varepsilon_y \cdot \frac{\bar{x}}{\bar{y}^2} \tag{1.16}$$

Hence, the absolute error is

$$\frac{\varepsilon_x}{\bar{y}} - \frac{\varepsilon_y \cdot \bar{x}}{\bar{y}^2} \tag{1.17}$$

and the relative error

$$\frac{\varepsilon_x}{\bar{x}} - \frac{\varepsilon_y}{\bar{y}} \tag{1.18}$$

Finally, we consider the error in taking some function of x such as square root, sine, etc.

If $x = \bar{x} + \varepsilon_x$ then, using Taylor series expansion

$$f(x) = f(\bar{x}) + \varepsilon_x f'(\bar{x}) + \frac{\varepsilon_x^2}{2!} f''(\bar{x} + \theta \varepsilon_x), \qquad (0 \leqslant \theta \leqslant 1) \tag{1.19}$$

Hence, the absolute error is approximately $\varepsilon_x f'(\bar{x})$ and the relative error $\varepsilon_x f'(\bar{x})/f(\bar{x})$ provided the second derivative of the function is not large. The propagation errors from the various operations can then be combined with the round-off errors discussed previously.

The precise calculation of the total accumulated error at any stage is very difficult to calculate for three reasons; some of the error estimates are based on neglecting products of errors, which will not be negligible as the errors grow. The size of the error may depend on the order in which the calculations are carried out. (See Example 1.6.) Also, there are cases where the size of the error depends on the size of the results and in these cases the most pessimistic result is assumed. Thus, it is possible to calculate an upper bound for the errors which, in almost all cases, is very much higher than the actual errors. It is also possible to make statistical estimates of the expected error but this is a rather more involved process which will not be pursued here. The reader who is in-

terested in obtaining further details on error analysis should refer to the books
of McCracken and Dorn (1964), Hamming (1962), or Ralston (1965). The latter
two books include some details on the statistical analysis of round-off error.

As an example of the calculation of the error, we will consider the
example $a/[b(c-d)]$ and introduce the reader to the graphical method of representing this process which is described, for example, in McCracken and Dorn
(1964).

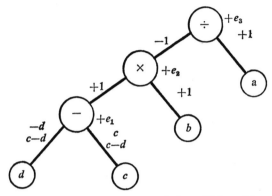

Fig. 1.1   Process graph for $a/[b(c-d)]$.

The graph shows the growth of the relative error if the calculation is done in the
order $r_1 = c - d$, $r_2 = b \times r_1$, and $r_3 = a/r_2$. The graph is used by starting at
the bottom and multiplying the accumulated error at any circle by the factor
alongside the line leading to the next circle. The round-off error is then added
to this. This total is then used as the accumulated relative error to be combined
with the appropriate multiplying factor at the next higher level as follows:

$$E_1 = \frac{\bar{c}}{\bar{c}-d}e_c - \frac{d}{\bar{c}-d}e_d + e_1$$

$$E_2 = E_1 + e_b + e_2$$

$$E_3 = -E_2 + e_a + e_3 \tag{1.20}$$

$$= \frac{-\bar{c}}{\bar{c}-d}e_c + \frac{d}{\bar{c}-d}e_d - e_b + e_a$$

$$-e_1 - e_2 + e_3$$

where

$$|e_1| \leqslant 2^{-t}$$

$$|e_2| \leqslant 2^{-t} \tag{1.21}$$

$$|e_3| \leqslant 2^{-t}$$

Note, that the bar notation refers to the machine number which is the rounded
version of the true answer.

1.5   Errors of computational methods

As well as the errors created by the arithmetic processes on a computer we must
also consider the errors generated by the approximation methods which are

used. A simple example is provided by the need to use functions such as sine, log, square root, etc., which are normally available in tabulated form. It is easy to see that the storage of this vast collection of numbers in a computer is not feasible, so other means must be sought. Fortunately, there are various methods of representing these functions by convergent series, which can be defined by a few simple rules which are easily programmed. A number of terms of the series are added together until the neglected error is acceptably small. Such an error due to the finite representation of an infinite process is called a truncation error.

The Taylor series representation of functions will be used to illustrate the use of convergent series, although this particular series is not suitable for the generation of function values in a computer. The general form of Taylor series with remainder is

$$f(x+h) = f(x) + hf'(x) + \frac{h^2}{2!}f''(x) + \cdots + \frac{h^{n-1}}{(n-1)!}f^{(n-1)}(x) + R_n \quad (1.22)$$

where

$$R_n = \frac{h^n}{n!}f^{(n)}(x+\theta h), \qquad 0 \leqslant \theta \leqslant 1 \tag{1.23}$$

It is valid in the interval $(x, x+h)$ provided the function is continuous and has continuous derivatives up to order n in the interval. This gives a convergent series if it can be shown that $\lim_{n \to \infty} R_n = 0$. In a computer calculation the terms are calculated one by one until R_n is so small that it is comparable in size to the computer round-off level. No further progress can then be made.

Consider

$$e^x = 1 + x + \frac{x^2}{2!} + \frac{x^3}{3!} + \frac{x^4}{4!} + \cdots$$

Let

$$A_0 = 1 \quad \text{and} \quad A_r = \frac{x \cdot A_{r-1}}{r}, \qquad r = 1, 2, \ldots \tag{1.24}$$

Then, the sums

$$S_0 = A_0 \qquad S_r = S_{r-1} + A_r, \qquad r = 1, 2, \ldots \tag{1.25}$$

give approximations to e^x. A simple set of instructions could thus be stored in the machine which will enable e^x to be calculated as required. In high-level languages these pieces of program are provided in the computer system but since these are programmed instructions they are slower than machine wired facilities, such as addition or multiplication.

The study of the convergence of series is outside the scope of this book but the reader should realize that, although the series e^x converges for all finite values of x, there are series where the sum of the terms increases without bound as n increases. The series for $(1-x)^{-1}$ is one such example for certain values of x

$$(1-x)^{-1} = 1 + x + x^2 + x^3 + \cdots \tag{1.26}$$

If the value $x = 1\cdot1$ is used in the expansion we have

$$1 + 1\cdot1 + 1\cdot21 + 1\cdot331 + \ldots \tag{1.27}$$

and it can be seen that each of the terms is greater than 1 so that the sum increases indefinitely.

Another approximation technique which is used for a variety of computational problems is the iteration method. This is a repetitive process which puts one approximation to the solution into a calculation, and uses this to produce a better approximation. When the sequence of results is converging a stage is reached where the successive values do not differ appreciably, and at this stage the calculation stops. It is, of course, important in an iteration procedure to have some means of halting the process when the results do not converge to a constant value. The precise definition of such a test is quite difficult.

The iteration technique in its simplest form can be applied to solve a single equation in one unknown, provided it can be put into the form $x = g(x)$ where $|g'(x)| < 1$ in the region of the iteration. The iteration then proceeds by taking an initial guess x_0 and calculating

$$x_1 = g(x_0)$$

This process is then repeated

$$x_{r+1} = g(x_r), \qquad r = 1, 2, \ldots \tag{1.28}$$

until the process has converged.

The importance of the derivative condition is easily shown by considering the example $x + 4\sqrt{(x)} - 8 = 0$ which has solutions $x = 29\cdot8552$ and $2\cdot1432$. We first consider the iteration

$$x = 8 - 4\sqrt{x} \tag{1.29}$$

starting from the first approximation $x_0 = 1$. The table of iterates is given by

$$
\begin{aligned}
x_0 &= 1 \\
x_1 &= 8 - 4 \times 1 = 4 \\
x_2 &= 8 - 4 \times 2 = 0 \\
x_3 &= 8 - 4 \times 0 = 8 \\
x_4 &= 8 - 4 \times 2\cdot828 = -3\cdot312
\end{aligned}
\tag{1.30}
$$

and no convergence is possible with this scheme since

$$g'(x) = \frac{-4}{2\sqrt{x}}$$

has the values

x	0	1	4
$g'(x)$	$-\infty$	-2	-1

$$\tag{1.31}$$

However, if the iteration scheme is written in the form

$$\sqrt{x} = \tfrac{1}{4}(8 - x) \quad \text{or} \quad x = \tfrac{1}{16}(8 - x)^2 \tag{1.32}$$

we find that $g'(x) = \frac{1}{16}(-16 + 2x)$ with values satisfying $|g'(x)| < 1$ in the range considered, i.e.,

x	0	1	2	4
$g'(x)$	-1	$-\frac{7}{8}$	$-\frac{3}{4}$	$-\frac{1}{2}$

(1.33)

The table of values is

$$
\begin{aligned}
x_0 &= 1 & x_6 &= 1\cdot9625 \\
x_1 &= 3\cdot0625 & x_7 &= 2\cdot2782 \\
x_2 &= 1\cdot5236 & x_8 &= 2\cdot0461 \\
x_3 &= 2\cdot6214 & x_9 &= 2\cdot2155 \\
x_4 &= 1\cdot8080 & x_{10} &= 2\cdot0912 \\
x_5 &= 2\cdot3963 & x_{11} &= 2\cdot1821
\end{aligned}
$$

(1.34)

and we see that the values are converging although the rate of convergence is slow. The value of $g'(x)$ at this point is $-0\cdot7321$ and, since this is fairly close to -1, the rate of convergence is slow.

1.6 Acceleration of convergence

In step-by-step processes, such as that described above, it is important to develop methods of speeding up the process. Since these ideas are used in several different problems, the basic principles of three acceleration methods will be discussed here, and applied to particular examples in later chapters. Firstly, we consider a simple iterative scheme $x_{n+1} = Kx_n + B$ and, as an example, take the values $K = \frac{1}{2}$ and $B = 1$. The values of successive iterates are plotted graphically in order to show that the process is converging to the value $x = 2$ and that this convergence is slow.

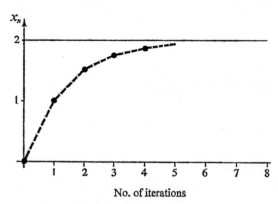

No. of iterations

Fig. 1.2   Iteration for $x_{n+1} = Kx_n + B$.

1.6.1 The over-relaxation method

We can see from the diagram that the distance from the true solution to the iterate is halved at each step. It would speed up the process if a bigger step towards the value 2 is taken at each stage. This is the motivation behind the method of over-relaxation which operates according to the followed equations. An intermediate value is formed

$$\tilde{x}_{n+1} = Kx_n + B \tag{1.35}$$

and then a corrected value is formed by taking the old value and adding on a multiple of the calculated increase, i.e.,

$$x_{n+1} = x_n + \omega(\tilde{x}_{n+1} - x_n) \tag{1.36}$$

which gives the equation

$$\begin{aligned} x_{n+1} &= x_n + \omega(K-1)x_n + \omega B \\ &= [\omega K + 1 - \omega]x_n + \omega B \end{aligned} \tag{1.37}$$

The value of ω is varied to achieve a faster rate of convergence and in certain cases, such as matrix iteration, it is possible to calculate the best value of ω theoretically. If $\omega > 1$ the system is said to be over-relaxed, whereas if $0 < \omega < 1$ the system is under-relaxed. Under-relaxation is not widely used but, in the solution of linear simultaneous equations, over-relaxation with $1 < \omega < 2$ is a very useful method. The effect can easily be seen by taking the previous numerical example with $\omega = \frac{3}{2}$. The successive iterates are now 0, $\frac{3}{2}$, $1\frac{7}{8}$, and $1\frac{31}{32}$ so that convergence in this case is twice as fast as for direct iteration. Indeed, in this example convergence takes place in one step if $\omega = 2$ but this results from the use of an artificially simple example. Some figures for over-relaxation used in conjunction with Newton's method are given in Example 1.9.

1.6.2 Aitken's Δ^2 process
The behaviour above (where the difference between the true value x^* and the iterate decreases in a fixed ratio) rarely occurs in practice but after a few initial iterations some problems give a close approximation to this. Assuming this fixed-ratio property is exact, a simple formula known as Aitken's Δ^2 process enables the final value of the converged solution to be calculated.

Assume

$$x^* - x_{n+1} = A(x^* - x_n) \tag{1.38}$$

Then

$$x^* - x_{n+2} = A(x^* - x_{n+1})$$

and dividing gives

$$\frac{x^* - x_{n+1}}{x^* - x_{n+2}} = \frac{x^* - x_n}{x^* - x_{n+1}} \tag{1.39}$$

$$(x^*)^2 - 2x_{n+1} \cdot x^* + x_{n+1}^2 = (x^*)^2 - (x_n + x_{n+2})x^* + x_n x_{n+2}$$

On eliminating the first term and rearranging

$$x^* = \frac{-x_{n+1}^2 + x_n x_{n+2}}{x_n - 2x_{n+1} + x_{n+2}} = x_n - \frac{(x_{n+1} - x_n)^2}{x_n - 2x_{n+1} + x_{n+2}}$$

$$= x_n - \frac{(\Delta x_n)^2}{\Delta^2 x_n} \tag{1.40}$$

The last expression uses the finite difference notation, which explains where the name of the process originates. The previous example will of course give the exact answer if this process is used, since it obeys the ratio rule exactly. Let $x_n = 0$, $x_{n+1} = 1$, and $x_{n+2} = 1\frac{1}{2}$.

Then, $\Delta x_n = 1$, $\Delta^2 x_n = -\frac{1}{2}$, and $x^* = 2$. A numerical example is given by Example 1.7.

1.6.3   Richardson's extrapolation method

This method is applied to processes which depend upon a step length h which can be chosen arbitrarily. This occurs, for example, in approximate integration where the interval is divided into strips, and also in the step-by-step integration of differential equations. The method can be used for processes in which the error can be expressed as a convergent power series in h, i.e., if the true value is Y and the approximation Y_1 then

$$Y = Y_1 + \sum_{j=k}^{\infty} a_j h_1^j \tag{1.41}$$

We assume that two calculations are performed with different values of h and the two results are then combined to eliminate the leading error term. The following equation

$$Y = Y_2 + \sum_{j=k}^{\infty} a_j h_2^j \tag{1.42}$$

and Eqn (1.41) are used. Multiply Eqn (1.41) by h_2^k and Eqn (1.42) by h_1^k and subtract, which gives an equation with the coefficient of h^k zero

$$Y[h_2^k - h_1^k] = h_2^k Y_1 - h_1^k Y_2 + \sum_{j=k+1}^{\infty} a_j [h_2^k h_1^j - h_1^k h_2^j]$$

Let h_2 be represented by a multiple of h_1 to simplify the notation, i.e., $h_2 = r \cdot h_1$. Then

$$Y = \frac{r^k Y_1 - Y_2}{r^k - 1} + \sum_{j=k+1}^{\infty} a_j \cdot h_1^k (r^k - r^j) \cdot h_1^j \tag{1.43}$$

The new estimate of the result is the first expression on the right-hand side and the error term now starts with a higher power of h. This process can be repeated with further values of h and removing more error terms but, in such a case, a careful investigation of the convergence of the process must be undertaken. This has been done for the method of Romberg integration which uses successive halving of the intervals, i.e., $r = \frac{1}{2}$, and the method is convergent for a wide class of functions. This method is discussed in Chapter 8 on approximate integration and Example 1.8 gives a table of figures for this process.

Bibliographic notes

Those books which contain material relevant to this chapter will be discussed first and then a survey will be given of books which cover most of the main topics of numerical analysis. Books concerned with specific topics will be discussed at the end of the relevant chapter. Titles are listed in the Bibliography on p. 183.

The books of Conte (1965), Fox and Mayers (1968), and McCracken and

Dorn (1964) all contain useful introductions to the subject of errors which are suitable for readers without specialist mathematical training. Conte also includes some details of the binary-number system and a summary of some of the mathematical topics which are needed for a study of numerical methods. Throughout the book by Fox and Mayers, considerable emphasis is placed on the analysis of errors, the groundwork for this being laid in Chapter 2 with a careful discussion of 'forward' and 'backward' error analysis. McCracken and Dorn give several examples of the use of the process graph to calculate the likely errors.

Ralston (1965) and Hamming (1962) also give details of the statistical approach to error analysis. This gives more realistic estimates of the errors but is not available to those with no training in statistical methods.

In the field of general books on computational methods, there are several books which have been available for some time and which laid the basis for later development. Many of these books were written when computation was frequently performed on a desk machine and they have some bias towards this form of computation. However, the books are quite detailed and contain methods which sometimes fall temporarily out of fashion. Many of the hints on computational difficulties in numerical methods are most valuable. The books by Buckingham (1957), Hildebrand (1956), Scarborough (1958), Householder (1953), and Todd (1962) are recommended, although considerable mathematical background is required, particularly for the last two texts. There are two useful books, Lanczos (1957) and Hamming (1962) written for the scientist and engineer with some mathematical knowledge who wishes to gain an insight into numerical methods. These books give considerable background of the motivation and usefulness of methods rather than mathematical theorems. Both of these books contain substantial sections on Fourier methods and associated work. The book by Lanczos has one feature which all readers will find valuable; namely, the introductions to the various chapters give concise surveys of the topics discussed which are both interesting and informative.

There are several books which contain material which could be introduced to undergraduates, including Bull (1966), Conte (1965), Fox and Mayers (1968), Henrici (1964), and McCracken and Dorn (1964). The text by Bull contains a wealth of numerical examples and ALGOL programs for the various algorithms. Very little mathematical background is given. The treatment of the subject and the topics chosen are biased towards the finite-difference calculus and the advanced methods are not discussed. The book by Conte contains several numerical examples and FORTRAN programs and provides several mathematical results with proofs. The book would form an excellent introduction to numerical methods for mathematicians. Henrici concentrates still more on the mathematical side. He gives a treatment more suitable as a basis for theoretical development rather than a starting point for practical calculations on a computer. Engineers and scientists will find the book by McCracken and Dorn easy to read and strongly orientated towards practical computation. The more advanced methods are not discussed but the simpler methods are well illustrated with numerical examples and FORTRAN programs. The book by Fox and Mayers contains computational details of many standard procedures and considerable discussion of the errors involved. The standard of mathematics required is higher than that for McCracken and Dorn but not as high as that required for the more specialized texts below.

A brief discussion of many standard methods is also given in *Modern Computing Methods* (1961).

Useful treatment of certain problems is provided in the texts edited by Ralston and Wilf, Volume 1 (1960) and Volume 2 (1967). For the problems considered a flow chart and program is provided together with detailed mathematical analysis of the method.

For those interested in a more advanced mathematical treatment of these subjects the books by Ralston (1965) and Isaacson and Keller (1966) are recommended. The book by Ralston has more emphasis on details of the computational methods while Isaacson and Keller are more concerned with the development of the mathematical theorems and rigorous investigation of the errors.

Worked examples

1 Convert the following numbers from binary to decimal: 10111 and 1010001.

$$10111 = 16 + 4 + 2 + 1 = 23$$

By nested multiplication

$$(((1 \times 2 + 0) \times 2 + 1) \times 2 + 1) \times 2 + 1 = 23$$
$$\quad\; 2, \qquad 4, \quad 5, \quad 10, \; 11, \; 22, \; 23$$

$$1010001 = 64 + 16 + 1 = 81$$

By nested multiplication

$$(((((1 \times 2 + 0) \times 2 + 1) \times 2 + 0) \times 2 + 0) \times 2 + 0) \times 2 + 1 = 81$$
$$\qquad 2, \quad 2, \quad 4, \quad 5, \quad 10, \; 10, \quad 20, \; 20, \quad 40, \; 40, \quad 80, \; 81$$

2 Convert the following numbers from decimal to binary: 28, 53. By subtracting the highest possible power of two at each stage

$$28 = 16 + 8 + 4 = 11100$$

By successive division by 2

2	28		
2	14	r	0
2	7	r	0
2	3	r	1
2	1	r	1
2	0	r	1

Answer 11100

By subtracting the highest power of two at each stage

$$53 = 32 + 16 + 4 + 1 = 110101$$

By successive division by 2

2	53		
2	26	r	1
2	13	r	0
2	6	r	1
2	3	r	0
2	1	r	1
2	0	r	1

Answer 110101

3 Perform the following calculations in binary arithmetic.

$$13 + 9 = 22, \qquad 25 - 12 = 13, \qquad 5 \times 7 = 35, \qquad 28 \div 4 = 7$$

```
    1101          11001           101            111
  + 1001         - 1100           111        100) 11100
  ------         ------          -----            100
   10110          1101            101            ----
                                  101             110
                                 -----            100
                                  101            ----
                                  101             100
                                 ------           100
                                 100011          ----
                                                  100
```

4 The method of subtraction by complement is a very useful and widely used method. This can be illustrated in decimal arithmetic. The subtraction of 273 is equivalent to subtracting $1000 - 727$, i.e., subtracting 1000 and adding 727.

$$278 - 273 = \begin{array}{r} 278 \\ 727 \\ \hline 1005 \end{array} \qquad \textbf{Answer} \quad 5$$

The complement is formed by subtracting the number from 999 and adding 1 to the units column. The subtraction is performed by adding the complement and then deducting 1 from the thousands column.

The same process can be done in binary and corresponds to very simple electronic circuitry. Subtract 3 from 7 in binary. The complement of 11 is 101. Add the complement to 111.

$$\begin{array}{r} 111 \\ 101 \\ \hline 1100 \end{array} \qquad \textbf{Answer} \quad 100_2 = 4$$

5 If a calculation involves the subtraction of two nearly equal numbers considerable errors can result. Consider the calculation $1/(0 \cdot 003\,146 - 0 \cdot 003\,130)$ carried out to three significant figures.

$$\frac{1}{0 \cdot 315 \times 10^{-2} - 0 \cdot 313 \times 10^{-2}} = \frac{1}{0 \cdot 2 \times 10^{-4}}$$
$$= 5 \cdot 00 \times 10^4$$

The correct answer is

$$\frac{1}{0 \cdot 16 \times 10^{-4}} = 6 \cdot 67 \times 10^4$$

so that an error of approximately 25% has been introduced.

This problem occurs in the solution of a quadratic equation if b^2 is very large compared to $4ac$. The two quantities b, $\sqrt{(b^2 - 4ac)}$ will be very nearly equal and one of the roots will be subject to large errors. For example, if b is positive this will be the root $x = (-b + \sqrt{(b^2 - 4ac)})/2a$. The solution is to find the other root, $x = (-b - \sqrt{(b^2 - 4ac)})/2a$, and the second root can then be found from the equation $x_1 . x_2 = c/a$. The equation $x^2 - 1000 \cdot 01x + 10 = 0$ has roots $x_1 = 1000$ and $x_2 = 0 \cdot 01$. Using the formula to six decimal places gives $x_2 = 0 \cdot 015$ which is an error of 50%. If the largest root is found first we have $x_1 = 999 \cdot 995$ and, using the equation $x_1 . x_2 = 10$, gives the value $x_2 = 0 \cdot 009\,999\,95$ and the errors are quite small.

6 A simple change in the order of calculations can sometimes have a significant effect as shown by the following examples. In the examples the arithmetic is performed to four significant figures. In computer terms this means that eight places are available for each calculation but the answer must then be rounded to four significant figures. A simple expression demonstrates this.

$$225 \cdot 1 - (224 \cdot 8 + 0 \cdot 1572)$$
$$= 225 \cdot 1 - (225 \cdot 0)$$
$$= 0 \cdot 1000$$

Alternatively, we have

$$225 \cdot 1 - 224 \cdot 8 - 0 \cdot 1572$$
$$= 0 \cdot 3000 - 0 \cdot 1572$$
$$= 0 \cdot 1428$$

These two addition sums show how the order of additions affect the answer.

991·1	0·5112
327·6	0·1001
1318·7	0·6113
1319·	1·543
225·0	2·1543
1544·	3·712
85·67	5·866
1629·67	25·54
1630·	31·406
75·61	31·41
1705·61	75·61
1706·	107·02
25·54	107·0
1731·54	85·67
1732·	192·67
3·712	192·7
1735·712	225·0
1736·	417·7
1·543	327·6
1737·543	745·3
1738·	991·1
0·1001	1736·4
1738·	1736
0·5112	
1738·5112	
1739·	

The correct solution is

0·5112
0·1001
1·543
3·712
25·54
75·61
85·67
225·0
327·6
991·1

1736·3863

7 An example in the next chapter shows how a quadratic equation $x^2 - x + 0·16 = 0$ can be solved by the iteration equation $x_{n+1} = x_n^2 + 0·16$. The starting value chosen is $x_0 = 0·15$ and the sequence of iterates is shown below. The first and second differences are calculated and Aitken's Δ^2 process is applied. The right-hand side shows that the values found by this process converge twice as fast as the raw values. The Δ^2 process is defined by the equation

$$x^*_{n+1} = x_{n+1} - \frac{(\Delta x_n)^2}{\Delta^2 x_n}$$

The figures are given to six decimal places in Table 1.1.

Table 1.1

x	Δx	$\Delta^2 x$	x^*
0·150 000	0·032 500	−0·021 650	0·198 686
0·182 500	0·010 805	−0·006 744	0·199 811
0·193 306	0·004 061	−0·002 475	0·199 969
0·197 367	0·001 586	−0·000 957	0·199 995
0·198 953	0·000 629	−0·000 379	
0·199 582	0·000 250	−0·000 150	
0·199 832	0·000 100	−0·000 060	
0·199 932	0·000 040	−0·000 024	
0·199 972	0·000 016	−0·000 009	
0·199 988	0·000 007		
0·199 995			

8 A simple approximation to the area under a curve is provided by using a straight-line approximation between successive points and using the trapezium rule to find the area as shown in Fig. 1.3.

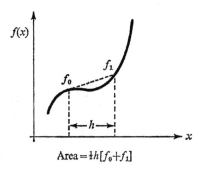

$$\text{Area} = \tfrac{1}{2}h[f_0 + f_1]$$

Fig. 1.3 The trapezoidal rule.

$$\text{Area} = \frac{h}{2}[f_0 + f_1]$$

Normally, the interval is split into several sub-intervals so that the actual formula used is

$$\text{Area} = \frac{h}{2}[f_0 + 2f_1 + 2f_2 + \cdots + f_n]$$

where h is the distance between successive function values. The following results in column one of Table 1.2 were obtained for the integration of x^4 between $x = 0$ and $x = 2$ for 4, 8, 16, and 32 intervals respectively.

Table 1.2

n	Integral	R_1	R_2	R_3
4	7·062 50			
		6·401 04		
8	6·566 41		6·399 99	
		6·400 06		6·400 00
16	6·441 65		6·400 00	
		6·400 01		
32	6·410 42			

The error formula for the trapezoidal integration rule has a form suitable for using Richardson extrapolation, i.e.,

$$I = I_n + \sum_{r=1}^{\infty} a_r h^{2r}$$

Hence,

$$I = I_{2n} + \sum_{r=1}^{\infty} a_r \left(\frac{h}{2}\right)^{2r}$$

$$4I = 4I_{2n} + a_1 h^2 + \sum_{r=2}^{\infty} 4a_r \left(\frac{h}{2}\right)^{2r}$$

Therefore, the term in h^2 can be eliminated by subtracting these two equations.

$$3I = 4I_{2n} - I_n + \sum_{r=2}^{\infty} b_r (h)^{2r}$$

The values in column R_1 are found by the formula

$$R_1 = \frac{4I_{2n} - I_n}{3}$$

The values in the subsequent columns are found by the rule $R_m = (2^{2m} I_{2n} - I_n)/(2^{2m} - 1)$ applied to each column. The final column therefore has an error term of order h^8. It can be seen that the values down any forward diagonal are converging to the correct value

$$6.4 = \int_0^2 x^4 \, dx = \left[\frac{x^5}{5}\right]_0^2$$

9 It is interesting to observe the effect of over-relaxation on Newton's method applied to the problem $x^{20} - 1$. Newton's method is described in the next section. A section of the results of Newton's method, starting from the initial approximation $x_0 = 10$, is given in Table 1.3. The subsequent columns of the table give a section of the results when over-relaxation is used with $K = 2.0$.

Table 1.3

Newton's method

Iteration	Result	Iteration	Result
1	9·500 000	38	1·423 993
2	9·025 000	39	1·352 854
3	8·573 750	40	1·285 372
4	8·145 063	41	1·221 527
5	7·737 809	42	1·161 567
6	7·350 919	43	1·106 394
7	6·983 373	44	1·058 397
8	6·634 204	45	1·022 485
9	6·302 494	46	1·004 132
10	5·987 369	47	1·000 158
11	5·688 001	48	1·000 000
12	5·403 601	49	1·000 000

Over-relaxation

Iteration	Result	Iteration	Result
1	9·000 000	13	2·541 866
2	8·100 000	14	2·287 679
3	7·290 000	15	2·058 911
4	6·561 000	16	1·853 020
5	5·904 900	17	1·667 719
6	5·314 410	18	1·500 953
7	4·782 969	19	1·350 902
8	4·304 672	20	1·216 142
9	3·874 205	21	1·096 956
10	3·486 784	22	1·004 495
11	3·138 106	23	0·995 877
12	2·824 295	24	1·004 456

It can be seen that the over-relaxation produces a much faster rate of convergence initially; the first 6 iterations make as much progress as 12 Newton iterations. However, when the iteration is near the root the Newton method rapidly converges whereas the over-relaxation method oscillates very slowly to convergence. From iteration number 46 the Newton method takes three more steps. From a similar value at iteration number 22 the over-relaxation method takes over 150 steps more.

If an intermediate value of the relaxation factor is taken, say $K = 1·5$, the iteration reaches the value 1·001 752 at iteration number 30 and then oscillates to convergence to six decimal places at iteration number 43. Clearly, the method of over-relaxation should not be used without careful analysis.

10 The following example also shows the care which should be exercised in interpreting

Table 1.4

Results 1	Results 2
0·880 112	0·880 112
0·880 112	0·880 112
0·880 112	0·880 112
0·880 112	0·880 112
0·880 112	0·880 112
0·880 112	0·880 112
0·880 112	0·880 112
0·880 111	0·880 112
0·880 108	0·880 112
0·880 094	0·880 114
0·880 022	0·880 124
0·879 662	0·880 171
0·877 867	0·880 407
0·868 945	0·881 587
0·825 675	0·887 514
0·639 579	0·917 753
0·178 369	1·085 118
0·000 301	2·507 456
0·000 000	165·202 101
	$>10^6$

computer results. The problem to be solved is to find the root of the equation

$$5x - 3x^5 = 0$$

using the iteration equation

$$x_{n+1} = \tfrac{3}{5}x_n^5$$

with a first approximation which, to six decimal places, is $x_0 = 0\cdot880\,112$. The computer results of Table 1.4 were obtained using exactly the same computer program in both cases and the same computing service.

    The reader will observe that the initial approximation gives a gradient of the iteration function equal to 5 so that the process is not expected to converge. The first few values do not diverge because the initial approximation is very nearly the correct value of the root. However, the rounding errors gradually reveal the true divergent nature of the process. In the one case the process converges to another root at $x = 0$ and in the second case the values grow without bound. The iteration diagram is shown in Fig. 1.4.

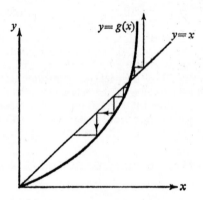

Fig. 1.4 A divergent iteration.

    The reason for the two different sets of results is that in the first case the results were obtained on an ICL 1904 with 37 bits in the mantissa. The second set of results were calculated on an ICL 1906A which uses multiple length for arithmetic calculations and rounds to 74 binary places within a calculation.

    Clearly, the process of transferring a program from one machine to another, even when these are theoretically compatible, may not be as simple as it seems.

Problems

1 Convert the following numbers to binary.

    37, 61, 211, 107, 57, 127

2 Convert the following binary numbers to decimal values.

    1101, 11110, 1010110, 11011011, 1010101, 11001100

3 Convert the following numbers to binary and carry out the calculations. Check the answers by converting back to decimal numbers.

    $23 + 17$, $18 - 12$, 15×14, $63 \div 9$

4 Derive a formula for the absolute error in adding four numbers, x_1, x_2, x_3, x_4, in succession assuming no error in the x_i. If the numbers are in order of size so that $x_1 > x_2 > x_3 > x_4$, state in which order the successive additions should be made. What formula is derived if the numbers are added in pairs and the resulting pair of numbers are added together?

2

Solution of nonlinear equations

There are some equations of a nonlinear type which will be familiar to readers because there are analytical methods leading to a formula for the solution. The solution of quadratic equations, or trigonometric equations such as $\sin 3x + 2\cos 3x = 2$, provide simple examples. However, many nonlinear equations cannot be solved directly by analytical methods and methods based on approximation must be used.

2.1 A simple iteration example

The method of simple iteration is easy to use and can be applied to a wide variety of problems. In its simplest form the iteration equation is obtained by rearranging the equation to isolate x on one side of the equation. An approximation to x is inserted on the other side of the equation and a new value of x is calculated. This new value of x is used in the calculation to give a further value of x and the process is repeated. A simple example of such an equation is given by Eqn (2.3).

If the method is to be successful these values must approach closer and closer to the true value of the solution. In such a case we say that the method is convergent. Conditions which will ensure convergence will be an important part of the discussion in this chapter. Unless careful analysis is used it is very easy to choose a method which gives a sequence of values which move further and further from the root.

As an example of the method we will study the quadratic

$$x^2 - x + 0.16 = 0 \tag{2.1}$$

This example is chosen because the solutions $x = 0.2$ and $x = 0.8$ are easily found. It is not intended that iteration should be seriously considered for the solution of quadratics! The equation can be written in the form

$$x = x^2 + 0.16 \tag{2.2}$$

Assume that a rough sketch of the function has been made which indicated that the smallest root was approximately 0.15. Substituting the value $x_0 = 0.15$ on the right-hand side we find a value $x_1 = 0.1825$. This value can then be inserted in the right-hand side and the process repeated according to the general formula

$$x_{n+1} = x_n^2 + 0.16, \quad n = 0, 1, \dots \tag{2.3}$$

The values obtained are given in Table 2.1.

Table 2.1

Simple iteration, $x_{n+1} = x_n^2 + 0.16$

x_0	0.150 000	x_6	0.199 832
x_1	0.182 500	x_7	0.199 932
x_2	0.193 306	x_8	0.199 972
x_3	0.197 367	x_9	0.199 988
x_4	0.198 953	x_{10}	0.199 995
x_5	0.199 582	x_{11}	0.199 999

We note that the process is converging to the true solution $x = 0.2$ but the rate of convergence is slow.

It is easily shown that the process does not always converge. Consider the equation

$$2x^2 - x = 0 \tag{2.4}$$

with roots $x = \frac{1}{2}$ and $x = 0$ and use the starting approximation $x_0 = 1$ with the iteration equation

$$x_{n+1} = 2x_n^2, \qquad n = 0, 1, \ldots \tag{2.5}$$

The sequence of approximations is $x_0 = 1$, $x_1 = 2$, $x_2 = 8$, and $x_3 = 128$ which is clearly not converging.

In this chapter we will be concerned with the problems of finding a good initial approximation and choosing an iterative scheme with good convergence properties.

2.2 Finding initial approximations

There are several useful properties of polynomials which enable approximations to the roots to be found and these are presented in Chapter 3. For a general nonlinear equation there are very few methods available. The following methods may be helpful.

1 It may well be that a knowledge of the physical circumstances of the problem will lead to a good approximation for the roots.
2 Alternatively, it is often possible to make a rough sketch of the function which will indicate the approximate position of the roots.
3 The equation can sometimes be split into two parts and the intersection of the graphs of the two functions may indicate where the roots lie more clearly than the graph of the original function.
4 If an automatic computer program is under consideration then calculation of function values could be carried out systematically until two values of opposite sign are found. For a continuous function it follows that these values enclose a root.
5 It is sometimes the case that in certain regions some part of the equation is negligible. It may then be possible to obtain an approximate root by solving the remaining part of the equation.

B

2.3 Simple iteration

The simple method used in the introductory section could be used for iteration if the formula was convergent. The general form of the method is obtained by rearranging the formula $f(x) = 0$ to give

$$x = g(x) \qquad (2.6)$$

leading to the iteration form

$$x_{n+1} = g(x_n), \qquad n = 0, 1, \ldots \qquad (2.7)$$

The first approximation x_0 is inserted in the right-hand side to give a new approximation x_1 on the left. This value is inserted in the right-hand side and the process repeated until the values have converged.

It is easy to show graphically the circumstances under which the process will converge. Figure 2.1(a) shows a typical graph in which the process is convergent. The curve $y = g(x)$ is drawn and this enables the value $y_0 = g(x_0)$ to be read from the graph. The value $x_1 = y_0$ is then required and this is found by drawing the line $y = x$, so that the appropriate x value can be read.

By drawing the lines leading from x_0 to x_1 etc., it can be seen that the values are converging to the point x^*, where $x^* = g(x^*)$, which is the solution of

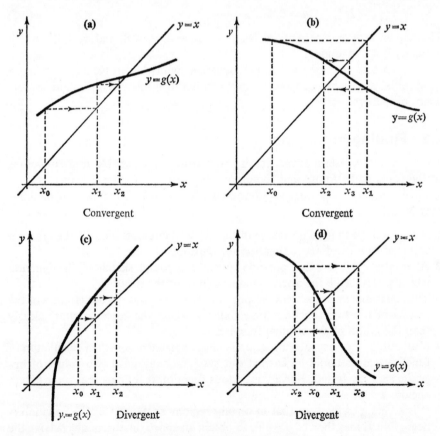

Fig. 2.1 Simple iteration.

the equation. In this type of convergence the difference between the approximation and the true solution always has the same sign so that the values steadily
approach the final limit. This type of behaviour is known as monotonic convergence.

Figure 2.1(b) shows another possible type of convergence where the
values oscillate on either side of the true solution. This oscillatory type of
convergence is very convenient because the latest two values give bounds within
which the true solution must lie. However, it is equally possible that the method
will diverge and Figs 2.1(c) and (d) show two simple examples of how this
might happen.

It is clear from studying the graph that the process will be convergent
when the gradient of the function $g(x)$ is less than that of the line $y = x$, i.e.,
in the region covered by the iteration we require

$$|g'(x)| < 1 \tag{2.8}$$

This can also be demonstrated theoretically. We have

$$x_{n+1} = g(x_n), \qquad n = 0, 1, \ldots \tag{2.9}$$

and the true solution x^* satisfies

$$x^* = g(x^*) \tag{2.10}$$

Subtracting these equations gives

$$x^* - x_{n+1} = g(x^*) - g(x_n) \tag{2.11}$$

Using the mean-value theorem the right-hand side becomes

$$g(x^*) - g(x_n) = (x^* - x_n)g'(\zeta) \tag{2.12}$$

where ζ is some value lying between x^* and x_n. If we define $\varepsilon_n = x^* - x_n$ the
above equation becomes

$$\varepsilon_{n+1} = g'(\zeta) . \varepsilon_n \tag{2.13}$$

and, if $|g'(\zeta)| < 1$, then the errors will always become smaller from step to step.
For $|g'(\zeta)| > 1$ the errors will grow.

2.4 Bisection method

The problem with the above method is that it may not be possible to choose any
iteration formula which satisfies the convergence condition (2.8). The method
of bisection has the advantages that it is always convergent and is very simple.
However, the rate of convergence is slow and when the iterations have ap-
proached close to the root it is usually desirable to use another method with
faster convergence. This is illustrated by Example 2.1.

It is required to find the values of x such that $f(x) = 0$. In order to
start the method, values of the function are calculated at a sequence of points
until two points x_0 and $\tilde{x}_0$ are found at which the function values have opposite
signs. Assuming the function is continuous, there must be a root between x_0 and
$\tilde{x}_0$. The function is then evaluated at the point $x_1 = \frac{1}{2}(x_0 + \tilde{x}_0)$. The point $\tilde{x}_1$ is
chosen as that point from the pair $x_0, \tilde{x}_0$ which has a function value opposite

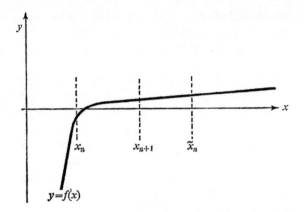

Fig. 2.2   Bisection method.

in sign to $f(x_1)$. An interval x_1 to $\tilde{x}_1$ has been obtained which still contains a root and which is half the size of the original interval. The process is then repeated using the formula $x_{n+1} = \frac{1}{2}(x_n + \tilde{x}_n)$ until the upper and lower bounds on the root are sufficiently close. Figure 2.2 shows a typical step of the method.

As well as the problem of slow convergence there are other limitations to the method. If there is a double root, or multiple root of even order, then the function will not change sign in the neighbourhood of the root and the method of bisection cannot be used. Also, the method is not suitable for finding complex roots.

2.5   Newton's method

The standard methods of calculus can be used in an attempt to find a method with a more satisfactory rate of convergence. If the iteration has reached the point x_n then an increment Δx_n is required which will take the process to the solution point x^*. If we expand $f(x^*)$ in Taylor series we have

$$0 = f(x^*) \equiv f(x_n + \Delta x_n) = f(x_n) + \Delta x_n f'(x_n) + \frac{(\Delta x_n)^2}{2!} f''(x_n) + \cdots \quad (2.14)$$

If the distance Δx_n between the present iteration point and the true solution is sufficiently small, then taking the first two terms on the right-hand side of Eqn (2.14) gives

$$0 \approx f(x_n) + \Delta x_n f'(x_n)$$

$$\Delta x_n \approx \frac{-f(x_n)}{f'(x_n)} \quad (2.15)$$

This leads to the formula known as Newton's method

$$x_{n+1} = x_n - \frac{f(x_n)}{f'(x_n)} \quad (2.16)$$

The method is shown graphically in Fig. 2.3. If a tangent to the curve is drawn at the point x_0 then

$$\tan\theta = \frac{f(x_0)}{-\Delta x} = f'(x_0) \quad (2.17)$$

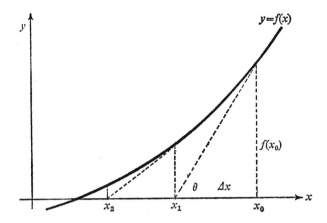

Fig. 2.3 Newton's method.

so that the step Δx can be found graphically by drawing a tangent to the curve at the current iteration point, and finding where it cuts the x axis. This is used as the next iteration point. The graphical approach is helpful in demonstrating the convergence properties of Newton's method.

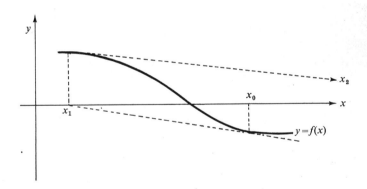

Fig. 2.4 Divergence of Newton's method.

There are various ways in which the method can fail to converge and a few sketches will show the kind of behaviour which can arise. Figure 2.4 shows an example of a curve which bends towards the axis, and it is clear that Newton's method is diverging. As a contrast, it is clear that for a curve such as in Fig. 2.3 Newton's method gives monotonic convergence.

Another possibility is shown by the oscillation of Fig. 2.5. Here, the first gradient at the point x_0 gives a point x_1 at which a second gradient is taken. The second gradient then generates a point in very much the same area as x_0 so that the process oscillates around a point which is not a root. (See also Example 2.2.) It is, of course, essential in an iteration process on a computer to have some means of stopping the process if no convergence is taking place. A count which stops the process after a given number of iterations is a simple method of achieving this. There are also problems which arise when $f'(x)$ is close to zero as in the case of multiple roots. This is demonstrated in Example 2.3.

In view of the difficulties which may arise when using Newton's method, care is needed to check that the process converges satisfactorily. Convergence can be guaranteed when the function has a second derivative which does not change sign in the region of the iteration, and which satisfies the following condition

$$f(x)f''(x) > 0 \qquad\qquad (2.18)$$

If the second derivative can easily be calculated then this condition can be used to check for convergence. If convergence cannot be calculated mathe-

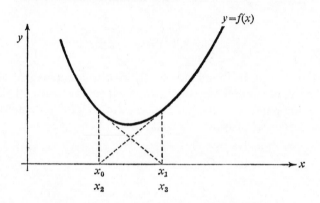

Fig. 2.5 Oscillation of Newton's method.

matically then it is possible, by careful programming, to monitor the progress of the iterations to see whether the iterations are converging. If there are no signs of convergence then it would be possible for the program to switch to some other iterative method with guaranteed convergence.

Even if the method is shown to be convergent there may be difficulties if the rate of convergence is slow. This is typified by the application of Newton's method to the equation $x^{20} - 1 = 0$. If the initial approximation $x_0 = \frac{1}{2}$ is used the following sequence of values is generated

$$x_1 = 26\,214\cdot9, \qquad x_2 = 24\,904\cdot1, \qquad x_3 = 23\,658\cdot9, \qquad x_4 = 22\,476\cdot0$$

The number of iterations before this monotonic sequence converges to the true value $x = 1$ will be very large. The reason for this is apparent if the Newton formula is written down

$$x_{n+1} = x_n - \frac{x_n^{20} - 1}{20x_n^{19}}$$

$$= \frac{19}{20}x_n + \frac{1}{20x_n^{19}} \qquad\qquad (2.19)$$

The second term is very small for $x > 2$ and the approximation $x_{n+1} = \frac{19}{20}x_n$ shows that the values decrease very slowly.

The attraction of Newton's method is that, when the errors are small, each error is proportional to the square of the previous error which gives more rapid convergence than the linear relationship which holds for simple iteration.

The error relationship for Newton's method is established by expanding $f(x)$ about the point x_n

$$0 = f(x^*) = f(x_n) + (x^* - x_n)f'(x_n) + \left(\frac{x^* - x_n}{2}\right)^2 f''(\zeta) \tag{2.20}$$

where ζ is some value of x between x_n and x^*. The term $f(x_n)/f'(x_n)$ in the Newton formula can be found from Eqn (2.16) and substituted in Eqn (2.20) to give

$$x_{n+1} = x_n + (x^* - x_n) + \frac{(x^* - x_n)^2}{2}\frac{f''(\zeta)}{f'(x_n)} \tag{2.21}$$

Thus, provided $f'(x_n)$ and $f''(\zeta)$ are nonzero, the error is a multiple of the square of the previous error.

$$x^* - x_{n+1} = -\left(\frac{x^* - x_n}{2}\right)^2 \frac{f''(\zeta)}{f'(x_n)} \tag{2.22}$$

2.6 Secant method

An obvious problem which may arise with Newton's method is the evaluation of the derivative $f'(x_n)$. For a function such as a polynomial this presents no problems since the nested-multiplication and synthetic-division algorithms give efficient automatic procedures for this. For most functions the derivative evaluation must be specially programmed and may take considerable computer time if the derivative function is complicated.

This problem can be avoided if an approximation to the derivative is used

$$f'(x_n) \approx \frac{f(x_n) - f(x_{n-1})}{x_n - x_{n-1}} \tag{2.23}$$

This leads to the formula for the secant method

$$\begin{aligned} x_{n+1} &= x_n - \frac{f(x_n)(x_n - x_{n-1})}{f(x_n) - f(x_{n-1})} \\ &= \frac{x_{n-1}f(x_n) - x_n f(x_{n-1})}{f(x_n) - f(x_{n-1})} \end{aligned} \tag{2.24}$$

The secant method must be started with two initial values x_0 and $\tilde{x}_0$. The values $f(x_0)$ and $f(\tilde{x}_0)$ are calculated which gives two points on the curve. It can be seen from the formula that the new point x_1 is found by linear interpolation. This is shown graphically in Fig. 2.6. The new point of the iteration sequence is the point at which the chord joining the two previous points cuts the x axis.

In the secant method the points are used in strict sequence. As each new point is found the lowest numbered point in the sequence is discarded. In this method it is quite possible for the sequence to diverge as shown by Fig. 2.6, where the point x_2 is clearly much further from the root than x_1. The rate of convergence of this method, when sufficiently close to the root, is superior to that of simple iteration but less than that of Newton's method. Each successive error depends on a power of the previous error in the following way.

$$e_{n+1} = K_n(e_n)^{(1+\sqrt{5})/2} \tag{2.25}$$

The text by Balfour and McTernan (1967) gives further discussion on rates of convergence.

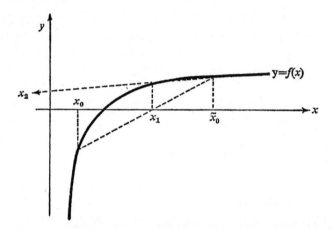

Fig. 2.6   Divergence of secant method.

2.7 Method of false position

A simple modification of the secant method will produce a method which is always convergent. If the choice of the two initial approximations can be made so that the two function values have opposite signs, then it is possible to generate a sequence of values which always possess this property. The value of x_1 is found by the linear interpolation formula and $f(x_1)$ is calculated. A value $\tilde{x}_1$ is chosen by taking the point from x_0 and $\tilde{x}_0$ at which the function value is opposite in sign to $f(x_1)$.

The values x_1 and $\tilde{x}_1$ now define a smaller interval in which a root must lie. The process is continued taking at each point the two values on opposite sides of the root. The convergence of the method of false position for the problem shown in Fig. 2.6 is shown in Fig. 2.7. The penalty involved in the modification of the method is that the high rate of convergence of the secant method is no longer applicable.

2.8 Comparison of methods

The simplest method is the method of bisection which is always convergent. It suffers from the disadvantages that the rate of convergence is slow and it cannot be used for multiple roots of even order. It is quite suitable as a preliminary method to find crude approximations to a root. These can then be refined by more sophisticated methods. Another method which will always converge is the method of false position. In certain circumstances this method is equivalent to the secant method and therefore has a good convergence rate.

However, Fig. 2.8 shows a situation which could result in quite slow convergence of the method of false position. When conditions for convergence are satisfied, then a method such as the secant method or Newton's method would be used.

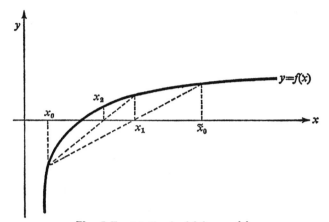

Fig. 2.7 Method of false position.

For an automatic computer program a combination of an always convergent method such as false position with a method which has fast convergence, such as Newton, would be desirable. The method of false position defines an interval in which the root lies, and the value given by Newton's method is accepted as the next iteration point if it is within this interval. If not, the value given by the method of false position is used as the next iteration point. By this means two values on opposite sides of the root will be maintained and Newton's method will only be used when it gives a value within this range.

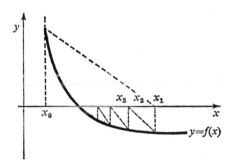

Fig. 2.8 Slow convergence of method of false position.

This approach is only suitable for roots of odd multiplicity, but when multiple roots of even order are involved there is always a region in which $f(x)f''(x) > 0$ so that Newton's method can be used with guaranteed convergence if a sufficiently close approximation can be found. There are also several methods of greater complexity which have higher rates of convergence than Newton's method. However, the same problem of investigating convergence exists and is made more difficult by the more complicated form of the higher-order formulae. An extensive discussion of other methods is given in Traub (1964).

2.9 Systems of nonlinear equations

A proper study of the solution of systems of equations is a topic requiring considerable analysis and is not suitable for study by the non-specialist. However, some of the methods of the previous section can be generalized to systems of equations, although the analysis of convergence properties is not simple.

As an example, Newton's method will be extended to the case of two simultaneous nonlinear equations in two unknowns

$$f(x, y) = 0$$
$$g(x, y) = 0 \tag{2.26}$$

Let x_0, y_0 be the first approximations to the solution and let $\Delta x, \Delta y$ be the increments needed to reach the true solution. Then, expanding in Taylor series as a function of two variables

$$0 = f(x_0 + \Delta x, y_0 + \Delta y)$$
$$= f(x_0, y_0) + \frac{\partial f}{\partial x}(x_0, y_0) \Delta x + \frac{\partial f}{\partial y}(x_0, y_0) \Delta y + \cdots$$
$$0 = g(x_0 + \Delta x, y_0 + \Delta y) \tag{2.27}$$
$$= g(x_0, y_0) + \frac{\partial g}{\partial x}(x_0, y_0) \Delta x + \frac{\partial g}{\partial y}(x_0, y_0) \Delta y + \cdots$$

If the second-order terms are ignored and the coefficients of $\Delta x, \Delta y$ are designated $\partial f/\partial x = a_{11}$, $\partial f/\partial y = a_{12}$, $\partial g/\partial x = a_{21}$, and $\partial g/\partial y = a_{22}$ then we have the two equations

$$a_{11} \Delta x_0 + a_{12} \Delta y_0 = -f(x_0, y_0)$$
$$a_{21} \Delta x_0 + a_{22} \Delta y_0 = -g(x_0, y_0) \tag{2.28}$$

These equations can be solved for the increments $\Delta x_0, \Delta y_0$ to give new approximations $x_1 = x_0 + \Delta x_0, y_1 = y_0 + \Delta y_0$. The process can then be repeated using the values x_1, y_1 on the right-hand side and in the partial derivatives to find the new values

$$x_{r+1} = x_r + \Delta x_r,$$
$$\qquad\qquad r = 0, 1, 2, \ldots \tag{2.29}$$
$$y_{r+1} = y_r + \Delta y_r,$$

In general terms we have a set of n functions in n variables

$$f^{(r)} = f_1[x_1^{(r)}, x_2^{(r)}, \ldots, x_n^{(r)}]$$
$$f_2[x_1^{(r)}, x_2^{(r)}, \ldots, x_n^{(r)}]$$
$$\cdots\cdots\cdots\cdots\cdots$$
$$f_n[x_1^{(r)}, x_2^{(r)}, \ldots, x_n^{(r)}] \tag{2.30}$$

A matrix $\mathbf{A}$ is formed with components

$$a_{ij} = \frac{\partial f_i}{\partial x_i} \tag{2.31}$$

and the increments $\Delta \mathbf{x}^{(r)}$ are found by solving successively the set of equations

$$A \, \Delta \mathbf{x}^{(r)} = -\mathbf{f}^{(r)}, \qquad r = 0, 1, \ldots \tag{2.32}$$

A simple numerical example of this type is given by Example 2.4.

Bibliographic notes

The standard texts such as McCracken and Dorn (1964), Conte (1965), and Bull (1966) contain descriptions of the simpler methods together with programs and examples. The text by Balfour and McTernan (1967) which specializes in this topic presents detailed treatment of the methods, with several numerical examples, at a level which is accessible to those without specialist mathematical training.

A more extensive treatment for those with further mathematical training is provided by Ralston (1965). Books which are suitable for the reader who wishes to make a detailed mathematical study of the subject are Wilkinson (1963), Traub (1964), and Ostrowski (1960).

Worked examples

1 This example shows the slowness of convergence which can arise for the method of bisection. Solve the equation $1/x + 1 = 0$ by bisection.
The initial values are

x	$-0{\cdot}5$	$-4{\cdot}0$
$f(x)$	$-1{\cdot}0$	$0{\cdot}75$

The sequence of iterates is

x	$-2{\cdot}25$	$-1{\cdot}375$	$-0{\cdot}9375$	$-1{\cdot}115\,625$	$-1{\cdot}046\,875$
$f(x)$	$0{\cdot}555\,556$	$0{\cdot}272\,728$	$-0{\cdot}066\,666$	$0{\cdot}135\,136$	$0{\cdot}044\,770$

x	$-0{\cdot}992\,187$	$-1{\cdot}019\,531$	$-1{\cdot}005\,859$	$-0{\cdot}999\,023$	$-1{\cdot}002\,929$
$f(x)$	$-0{\cdot}007\,874$	$0{\cdot}019\,157$	$0{\cdot}005\,825$	$-0{\cdot}000\,977$	$0{\cdot}002\,921$

x	$-0{\cdot}999\,515$	$-1{\cdot}001\,222$	$-1{\cdot}000\,368$
$f(x)$	$-0{\cdot}000\,485$	$0{\cdot}001\,221$	$0{\cdot}000\,368$

The figures obtained from Newton's method are

x	$-0{\cdot}75$	$-0{\cdot}937\,500$	$-0{\cdot}996\,094$	$-0{\cdot}999\,850$	$-1{\cdot}000\,000$
$f(x)$	$-0{\cdot}333\,333$	$-0{\cdot}066\,666$	$-0{\cdot}003\,921$	$-0{\cdot}000\,150$	$0{\cdot}000\,000$

The rate of convergence is much increased in the Newton iteration.

2 The problem of oscillation which can occur with Newton's method is shown very clearly by considering the application of the method to the equation

$$x^4 - 8x^3 + 23x^2 - 28x + 187/16 = 0$$

In fact, the equation has two roots near $x = 1$ and $x = 3$ but starting from the initial approximation $x_0 = 1\tfrac{1}{2}$ we have

$$x_1 = 1\tfrac{1}{2} - f(1\tfrac{1}{2})/f'(1\tfrac{1}{2}) = 2\tfrac{1}{2}$$
$$x_2 = 2\tfrac{1}{2} - f(2\tfrac{1}{2})/f'(2\tfrac{1}{2}) = 1\tfrac{1}{2}$$

and the process oscillates between the values $1\tfrac{1}{2}$ and $2\tfrac{1}{2}$.

3 When a multiple root is present there are difficulties in getting a close approximation to the root using Newton's method. This is because the evaluation of the polynomial $f(x)$ near the root almost always involves the subtraction of two nearly equal numbers. The error resulting from this is then magnified by division by a value of $f'(x)$ which is close to zero owing to the multiplicity of the root. The following table is the result of calculations carried out to six decimal places to find a root of the polynomial

$$f(x) = (x-1)^2(x-2)(x+3) = x^4 - x^3 - 7x^2 + 13x - 6$$
$$f'(x) = 4x^3 - 3x^2 - 14x + 13$$

x	$f(x)$	$f'(x)$	Δx
0·5	1·3125	5·75	0·228 260
0·728 260	−0·350 116	2·758 241	0·126 934
0·855 194	−0·092 544	1·335 022	0·069 320
0·924 514	−0·024 050	0·653 451	0·036 810
0·961 324	−0·006 155	0·322 639	0·019 077
0·980 401	−0·001 558	0·160 219	0·009 724
0·990 125	−0·000 393	0·079 874	0·004 920
0·995 045	−0·000 098	0·039 861	0·002 458
0·997 503	−0·000 024	0·020 032	0·001 198
0·998 701	−0·000 007	0·010 407	0·000 672
0·999 373	−0·000 001	0·005 019	0·000 199
0·999 572	−0·000 000	0·003 425	

Here, there are possible errors of 5×10^{-7} which are divided by approximately 5×10^{-4}. This introduces errors which could be as large as 10^{-4}. Hence, even the third decimal place may not be reliable and the fourth almost certainly is not. This is confirmed by a knowledge of the true value of $x = 1·0$. The result is not unexpected when it is realized that the function is effectively zero to six decimal places over the range $0·999\,500 \leqslant x \leqslant 1·000\,501$.

4 The method of solution of a system of nonlinear equations will be shown by a simple example with two equations in two unknowns. This will enable the reader to follow the process quite easily.

The equations below will be solved by Newton's method starting with an initial approximation $(x_0, y_0) = (3, 4)$

$$f(x, y) = x^2 + y^2 - 25 = 0$$
$$g(x, y) = x^2 - y^2 - 7 = 0$$

The equations for Newton's method applied to these equations are

$$2x\,\Delta x_r + 2y\,\Delta y_r = -f(x_r, y_r)$$
$$2x\,\Delta x_r - 2y\,\Delta y_r = -g(x_r, y_r)$$

Hence,

$$6\Delta x_0 + 8\Delta y_0 = 0$$
$$6\Delta x_0 - 8\Delta y_0 = +14$$
$$\Delta x_0 = +\tfrac{14}{12} = 1·166\,667$$
$$\Delta y_0 = -\tfrac{14}{12} = -1·166\,667$$
$$x_1 = 4·166\,667, \qquad y_1 = 2·833\,335$$

and

$$8·333\,334\,\Delta x_1 + 5·666\,670\,\Delta y_1 = -0·388\,887$$
$$8·333\,334\,\Delta x_1 - 5·666\,670\,\Delta y_1 = -2·333\,335$$
$$\Delta x_1 = -0·163\,333, \qquad \Delta y_1 = 0·171\,569$$
$$x_2 = 4·003\,334, \qquad y_2 = 3·004\,904$$

and

$$8{\cdot}006\,668\,\Delta x_2 + 6{\cdot}009\,808\,\Delta y_2 = -0{\cdot}056\,131$$
$$8{\cdot}006\,668\,\Delta x_2 - 6{\cdot}009\,808\,\Delta y_2 = 0{\cdot}002\,765$$

$$\Delta x_3 = -0{\cdot}003\,160, \qquad \Delta y_3 = -0{\cdot}004\,900$$

Therefore,

$$x_4 = 4{\cdot}000\,174, \qquad y_4 = 3{\cdot}000\,004$$

and we can see that the process is converging to one of the correct solutions.

Problems

1 The equation $2x^2 - x = 0$ is to be solved by simple iteration,

$$x_{n+1} = 2x_n^2$$

The derivative of $g(x)$ therefore equals $4x$. By mistake the first approximation is taken at

$$x_0 = \tfrac{1}{2}$$

where the derivative is

$$4.\tfrac{1}{2} = 2$$

therefore, the process should not converge. But successive values are $\tfrac{1}{2}, \tfrac{1}{2}, \ldots$. Explain this by reference to Fig. 2.1(c).

2 Find a root of the equation

$$x = -\log x$$

by simple iteration with an initial approximation

$$x_0 = 1{\cdot}0$$

What form should the iteration take in order to satisfy the convergence conditions?

3 Find the least positive root of

$$\tan y = e^y$$

4 A root of the equation

$$2x^3 + x^2 - 20x + 12 = 0$$

lies between $x = 0$ and $x = 1$. Use the method of false position to obtain a first approximation and then use Newton's method to improve the approximation.

5 Use Newton's method with an initial approximation of $x = 0{\cdot}0$. on the equation

$$x^3 - 6{\cdot}5x^2 + 14{\cdot}0x - 10{\cdot}0 = 0$$

Let the iteration terminate when successive values of x_r differ by less than 10^{-10}. The correct answer is $x = 2{\cdot}0$. If your answer differs significantly from this, give reasons for the discrepancy. Would the method of false position give better results? (The computer used must have a mantissa of at least 37 binary digits.)

3

Solution of polynomial equations

3.1 Introduction

In this chapter we are concerned with finding the roots of the polynomial

$$f(z) = a_0 + a_1 z + \cdots + a_n z^n, \qquad a_n \neq 0$$
$$\equiv a_n(z - z_1)(z - z_2) \cdots (z - z_n) \tag{3.1}$$

where the a_r are real coefficients and the z_r $(r = 1, 2, \ldots, n)$ are, in general, complex roots. A number α is called a root of the equation if $f(\alpha) = 0$. The simpler methods will be concerned with finding the real roots only, but for a complete solution all the roots must be found. Texts on algebra demonstrate that there are exactly n roots for a polynomial of degree n. It should be borne in mind that the complex roots occur in complex-conjugate pairs so that for any root $a + ib$ there corresponds the conjugate root $a - ib$.

At first glance there seems no reason to suppose there would be any particular difficulty in finding the roots of a polynomial. Analytically, a polynomial is a particularly simple function since all its derivatives are continuous in any finite region and can easily be integrated. However, the problem of computing accurate approximations to all the zeros of a polynomial can be extremely difficult, even for polynomials of degree as low as 20. The problems are particularly acute when multiple roots are present.

An example, given by Wilkinson (1963), illustrates how a very small variation in the coefficients of a polynomial can cause a large variation in some of the zeros. The following polynomial with zeros $z_k = k$ seems straightforward at first sight

$$f(z) = (z - 1)(z - 2) \cdots (z - 20)$$
$$= z^{20} - 210 z^{19} + 20615 z^{18} + \cdots + 20! \tag{3.2}$$

However, a change as small as 2^{-23} in the coefficient -210, which represents a small change in the tenth significant decimal figure, changes the roots $z_{10}, \ldots, z_{19}$ to complex roots. There are also significant changes in the values, for example,

	Old value	New values	
z_{14}	14	$13 \cdot 992 + 2 \cdot 519 i$	
z_{15}	15	$13 \cdot 992 - 2 \cdot 519 i$	(3.3)

The complete solution of a polynomial equation can be a most demanding task and it is recommended that any general-purpose program should use multi-length arithmetic to minimize the errors in difficult cases.

3.2 Polynomial arithmetic

One of the advantages of computation with polynomials is that there are simple algorithms for executing many of the typical calculations required with polynomials. We will consider the problems of finding, at any given point α, the value of the polynomial, its derivative, and the remainder on dividing by $z - \alpha$.

3.2.1   Nested multiplication

In the discussion on computer arithmetic we pointed out that addition and multiplication were usually basic computer operations, but other functions were provided by pieces of computer program which were slower. This is true for operations of the type z^r and it is desirable that this power, or exponentiation, operation be avoided. A polynomial evaluation calculated directly would require $n-1$ exponentiations, n multiplications, and n additions. The method of nested multiplication shown here requires only n multiplications and n additions.

The method in algebraic form applied to a cubic is

$$((a_3 z + a_2)z + a_1)z + a_0 \tag{3.4}$$

and the reader will see that the effect of this is to multiply a_3 by z^3 etc. This can be written in algorithmic form as follows

$$b_n = a_n$$
$$b_r = z b_{r+1} + a_r, \qquad r = n-1, n-2, \ldots, 1, 0 \tag{3.5}$$

which is simple to program on a computer. The quantity b_0 gives the value of the polynomial for a given value of z. As an example consider the value of the cubic $2z^3 - z^2 + 6$ at the point $z = 1\cdot1$.

$$
\begin{aligned}
b_3 &= 2\cdot0 \\
b_2 &= 1\cdot1 \times 2\cdot0 - 1\cdot0 = 1\cdot2 \\
b_1 &= 1\cdot1 \times 1\cdot2 + 0 = 1\cdot32 \\
b_0 &= 1\cdot1 \times 1\cdot32 + 6\cdot0 = 7\cdot452
\end{aligned}
\tag{3.6}
$$

This method should always be used for computer or desk-machine calculations.

3.2.2   Synthetic division

The process of division of a polynomial by a factor $z - \alpha$ is important for two reasons. It forms one component of the scheme for finding all roots of a polynomial and also, it enables the remainder theorem of algebra to be efficiently employed in computer calculation. The remainder theorem is developed by writing a polynomial in the form

$$f_n(z) = (z - \alpha)f_{n-1}(z) + R \tag{3.7}$$

The subscript denotes the degree of the polynomial and we see that the division will give a quotient of degree $n-1$ and a remainder R which is a constant. If we then let $z = \alpha$ we see that $R = f_n(\alpha)$ so that the remainder gives us the value of the function at $z = \alpha$.

We first show a typical long division by $z - \alpha$ and then the abbreviated table which gives the scheme known as synthetic division.

$$
\begin{array}{r}
z-\alpha \overline{)a_3z^3 + a_2z^2 + a_1z + a_0} \quad (a_3z^2 + (a_2 + a_3\alpha)z + a_1 + (a_2 + a_3\alpha)\alpha \\
\underline{a_3z^3 - a_3\alpha z^2}
\end{array}
$$

$$
\begin{array}{r}
(a_2 + a_3\alpha)z^2 + a_1z \\
\underline{(a_2 + a_3\alpha)z^2 - (a_2 + a_3\alpha)\alpha z}
\end{array} \qquad (3.8)
$$

$$
\begin{array}{r}
[a_1 + (a_2 + a_3\alpha)\alpha]z + a_0 \\
\underline{[a_1 + (a_2 + a_3\alpha)\alpha]z - [a_1 + (a_2 + a_3\alpha)\alpha]\alpha}
\end{array}
$$

$$
a_0 + [a_1 + (a_2 + a_3\alpha)\alpha]\alpha
$$

It can be seen that writing down the powers of z is superfluous, providing the coefficients are kept in the correct columns, and that the left-hand columns which produce zeros on subtraction can be ignored. A further simplification results if the subtraction which is performed is achieved by changing the sign of α and adding the two quantities. The following table is then produced.

$+\alpha$ $+$	a_3	a_2	a_1	a_0	
	0	$p_3\alpha$	$p_2\alpha$	$p_1\alpha$	(3.9)
	$p_3 = a_3$	$p_2 = a_2 + p_3\alpha$	$p_1 = a_1 + p_2\alpha$	$p_0 = a_0 + p_1\alpha$	

The lowest element of the last column of the scheme gives the remainder, i.e., the value of the polynomial at $z = a$. The coefficients of the quotient polynomial are given by p_3, p_2, and p_1.

The preceding process has the same computational steps as the nested-multiplication scheme but this scheme can also be extended to division by a quadratic factor, a process used when a conjugate-complex pair of roots occur in the solution of a polynomial. The synthetic division by a quadratic factor $z^2 + \alpha z + \beta$ has the form

	a_4	a_3	a_2	a_1	a_0
$-\alpha$		$-p_4\alpha$	$-p_3\alpha$	$-p_2\alpha$	
$-\beta$			$-p_4\beta$	$-p_3\beta$	$-p_2\beta$
	$p_4 = a_4$	$p_3 = a_3 - p_4\alpha$	$p_2 = a_2 - p_3\alpha - p_4\beta$	$p_1 = a_1 - p_2\alpha - p_3\beta$	$p_0 = a_0 - p_2\beta$

$$(3.10)$$

In this case, the remainder term is $p_1z + p_0$ and the coefficients of the quotient polynomial p_4, p_3, and p_2. The reader should check this process by carrying out the long-division process. A numerical example is given by Example 3.1.

3.2.3 Evaluation of the derivative

We consider the formulation of Eqn (3.7) showing the divisions by $z - \alpha$, i.e.,

$$
f_n(z) = (z - \alpha)f_{n-1}(z) + R \qquad (3.11)
$$

and differentiate both sides remembering that the remainder R is a constant.

$$
f'_n(z) = f_{n-1}(z) + (z - \alpha)f'_{n-1}(z) \qquad (3.12)
$$

By substituting the value of $z = \alpha$ we find that the second term on the right-hand side is zero and the derivative is given by the value $f_{n-1}(\alpha)$. This polynomial is available from the synthetic-division process and can be evaluated at the point $z = \alpha$ by a further synthetic division, or nested multiplication. The processes described in this section make the use of Newton's method for a polynomial particularly simple since the process can be used to evaluate $f(x)$ and $f'(x)$. A numerical example of this process is included in Example 3.1.

3.3 Finding initial approximations

If it is possible to do some preliminary work on an equation, before putting the data into the computer program, then methods can be used which involve the skill and intuition of the problem solver. The obvious method is to sketch the graph of the polynomial and, if possible, estimate the position of the roots. (This was discussed in Chapter 2.) There are various properties associated with polynomials which give help in estimating the approximate value of the roots, and some of these are presented here.

3.3.1 Useful properties of a polynomial
The polynomial can be represented in either of the two forms

$$f(z) = a_n z^n + a_{n-1} z^{n-1} + \cdots + a_0 \tag{3.13}$$

or

$$f(z) = a_n(z - z_1)(z - z_2) \cdots (z - z_n) \tag{3.14}$$

where z_r $(r = 1, 2, \ldots, n)$ are the roots of the equation. If the a_r $(r = 0, 1, 2, \ldots, n)$ are real, then all the roots are either real or they occur in complex-conjugate pairs. Thus, if there is a root $z = a + ib$ there is also the conjugate-complex root $z = a - ib$.

Comparison of the formulae of Eqns (3.13) and (3.14) give relationships between the coefficients a_r and the various symmetric products of the roots. The three which are of most interest are

$$\sum_{r=1}^{n} z_r = \frac{-a_{n-1}}{a_n} \tag{3.15}$$

$$\sum_{r=1}^{n} \sum_{s=r+1}^{n} z_r z_s = \frac{a_{n-2}}{a_n} \tag{3.16}$$

and

$$\prod_{r=1}^{n} z_r = (-1)^n \frac{a_0}{a_n} \tag{3.17}$$

The first two equations can be used to give an upper bound for the roots of the polynomial if the roots are all real.

We note that the maximum root x_{max} satisfies

$$x_{max}^2 \leqslant x_1^2 + x_2^2 + \cdots + x_n^2$$
$$= (x_1 + x_2 + \cdots + x_n)^2 - 2(x_1 x_2 + x_1 x_3 + \cdots + x_1 x_n + x_2 x_3 + \cdots)$$
$$= \left(\frac{a_{n-1}}{a_n}\right)^2 - 2\frac{a_{n-2}}{a_n}$$

Hence,

$$|x_{max}| \leqslant \sqrt{\left[\left(\frac{a_{n-1}}{a_n}\right)^2 - 2\frac{a_{n-2}}{a_n}\right]} \tag{3.18}$$

It is useful to be able to check for the case where a root is multiple since this may suggest the use of special methods. This can easily be done by noting if the root is also a root of successive derivatives of the polynomial. Let α be a root of multiplicity k, i.e.,

$$f(x) = (x - \alpha)^k g(x) \tag{3.19}$$

Then, differentiating $f(x)$ we see that α is also a root of $f'(x)$.

$$f'(x) = k(x - \alpha)^{k-1} g(x) + (x - \alpha)^k g'(x) \tag{3.20}$$

By further differentiation we see that all derivatives up to $f^{k-1}(x)$ also have a root at $x = \alpha$ and this enables the order of multiplicity of a root to be found.

There is a very simple rule, known as Descartes' rule of signs, which sometimes gives useful information on the position of roots. The method suffers the disadvantage that it may sometimes give no information at all. If all the coefficients of Eqn (3.13) are real then the number of positive roots of $f(z)$ equals the number of changes of sign in the sequence, $a_0, a_1, \ldots, a_n$, or is less than this by an even number.

The number of negative roots is related in a similar fashion to the polynomial produced by inserting the value $-x$ in the polynomial $f(x)$. The number of negative roots of $f(x)$ equals the number of sign changes in the polynomial $f(-x)$ or is less than this by an even number. Thus, we can see that if there are no sign changes in $f(x)$ then there are no positive roots. However, when the coefficients change sign several times, very little of value can be discovered.

3.3.2 Sturm sequences

The mathematical details of Sturm sequence theory are presented in Ralston (1965). It will suffice here to present the computational details of the use of Sturm sequences for finding the roots of polynomials.

In the simple case a sequence of polynomials are produced as follows: $f_0(x)$ is the original polynomial and $f_1(x)$ is the derivative of this polynomial. Subsequent polynomials are defined as the negative of the remainder obtained when dividing $f_r(x)$ by $f_{r+1}(x)$.

$$
\begin{aligned}
f_0(x) &= f(x) \\
f_1(x) &= f'(x) \\
f_0(x) &= f_1(x) q_1(x) - f_2(x) \\
f_1(x) &= f_2(x) q_2(x) - f_3(x) \\
&\cdots\cdots\cdots\cdots\cdots
\end{aligned}
\tag{3.21}
$$

If there are no repeated roots the sequence will terminate with a polynomial $f_k(x)$ ($k = n$) which is a constant. When repeated roots are present the process terminates with a polynomial $f_k(x)$ ($k < n$) and the value $n - k + 1$ represents the multiplicity of the root.

Assume, that there are no repeated roots; two values, a and b ($a < b$) are chosen and the values of the functions in the Sturm sequence evaluated at both points. Let $N(a)$ be the number of sign changes in the sequence $f_0(a), f_1(a)$, $\ldots, f_k(a)$, and $N(b)$ be the number of sign changes in $f_0(b), f_1(b), \ldots, f_k(b)$. The number of roots between a and b is given by $N(a) - N(b)$. It is assumed that there are no roots at a or b. If a point is chosen which is a root of the function, this value can be recorded immediately and the Sturm sequence evaluated at a different point.

A further complication can arise if there is no remainder at some point in the sequence of functions. In such a case a repeated root exists and this is given by the last member of the sequence $f_k(x)$ $(k < n)$. The Sturm sequence theory can still be used. A new sequence can be formed dividing by the common factor $f_k(x)$.

$$\tilde{f}_r(x) = \frac{f_r(x)}{f_k(x)}, \qquad r = 0, 1, \ldots, k \tag{3.22}$$

The pattern of sign changes can then be used as before to find the position of the other roots. Numerical examples of the use of Sturm sequences are given at the end of the chapter in Examples 3.3 and 3.4.

3.4 Complete solution of a polynomial

3.4.1 Deflation procedure

The method of solution is trivial to describe but to achieve accuracy of computation may be anything but trivial. We know that the polynomial has n roots, some of which are real and the others being conjugate-complex numbers. The complex roots can, therefore, be represented by a quadratic factor with real coefficients. That is,

$$(z - \alpha - i\beta)(z - \alpha + i\beta) = (z^2 - 2\alpha z + \alpha^2 + \beta^2) \tag{3.23}$$

First, it is necessary to do preliminary studies in order to find approximations to the roots by the methods described previously in Chapter 2. A suitable iterative method is then used to find an accurate value for the root.

A problem immediately arises since it frequently happens that iteration does not converge to the desired root. Also, in some cases one root dominates the process so that iterative processes converge to this root from a wide range of starting values. In this case, it is necessary to remove this root before iteration can be used to find further roots. This is done by the deflation process which uses synthetic division to divide by the accurate factor, which has been found, and gives as the quotient a new polynomial with one root less.

If the roots are complex it is usually more convenient to use Bairstow's iterative method for a quadratic factor since this avoids the use of complex numbers. When a quadratic factor has been found accurately, the synthetic-division process can again be used and this will reduce the degree of the polynomial by 2. This process of finding a root followed by deflation is continued until all the roots have been found. Examples 3.1 and 3.2 show the deflation process for a linear and quadratic factor, respectively.

This sequence of deflations and iterations can unfortunately generate large errors in the roots found later in the process. These errors can be minimized in two ways. Firstly, the roots should be found in order of increasing magnitude if possible. A discussion of this requirement together with an informative treatment of the problem of finding polynomial roots is given in

Bareiss (1967). Secondly, it is possible to invert the polynomial and then find the new roots in increasing order of magnitude which corresponds to decreasing order of magnitude for the original set. If we have a polynomial

$$f(z) = a_0 + a_1 z + a_2 z^2 + \cdots + a_n z^n \qquad (3.24)$$

form a new function by the transformation $z = 1/z$

$$f(z) = a_0 + \frac{a_1}{z} + \frac{a_2}{z^2} + \cdots + \frac{a_n}{z^n}$$

$$= \frac{1}{z^n}[a_0 z^n + a_1 z^{n-1} + \cdots + a_n z + a_n] \qquad (3.25)$$

The polynomial inside the brackets has roots which are the reciprocals of the original roots and the ordering by size is reversed.

In the first process the smallest roots will be the most accurate and in the second case the inverses of the largest roots will be most accurate.

If the values of $f(z)$ are tabulated for both sets of results the most accurate values from either set can be taken. Table 3.1 shows typical results. The entries in the table show the values of $f(z)$ for both the standard polynomial, labelled S, and the inverted polynomial, labelled I; the values should, of course, be zero if the roots are absolutely accurate. The symbol z_r $(r = 1, 2, \ldots 7)$ is included in the section of the table which is most accurate for each root.

Table 3.1

Errors in solving original and inverse polynomial

	z_1	z_2	z_3	
S	$5 \cdot 12 \times 10^{-12}$	$1 \cdot 28 \times 10^{-11}$	$8 \cdot 31 \times 10^{-11}$	
I	$2 \cdot 81 \times 10^{-8}$	$8 \cdot 93 \times 10^{-9}$	$6 \cdot 15 \times 10^{-9}$	

	z_4	z_5	z_6	z_7
S	$9 \cdot 23 \times 10^{-10}$	$1 \cdot 83 \times 10^{-9}$	$4 \cdot 81 \times 10^{-9}$	$2 \cdot 13 \times 10^{-8}$
I	$4 \cdot 51 \times 10^{-10}$	$3 \cdot 56 \times 10^{-10}$	$9 \cdot 05 \times 10^{-11}$	$8 \cdot 54 \times 10^{-11}$

We now consider the choice of iterative methods suitable for polynomials. Since, the derivative of a polynomial of degree n has degree $n - 1$, with $n - 1$ roots, it can be seen that a polynomial has a large number of turning points as demonstrated by the following graph in Fig. 3.1:

$$x^8 - 4x^7 - 46x^6 + 168x^5 + 553x^4 - 1988x^3 - 988x^2 + 5184x - 2880$$
$$= (x + 5)(x + 4)(x + 2)(x - 1)^2(x - 3)(x - 4)(x - 6)$$

A method such as Newton's method converges quickly in the neighbourhood of the root but is subject to convergence problems in the case of a highly oscillatory function such as a high-order polynomial. It is more desirable when dealing with a polynomial to use a method which is sure to converge, even if this means some

Graph of $p(x)=(x+5)(x+4)(x+2)(x-1)^2(x-3)(x-4)(x-6)$
$=x^8-4x^7-46x^6+168x^5+553x^4-1988x^3-988x^2+5184x-2880$

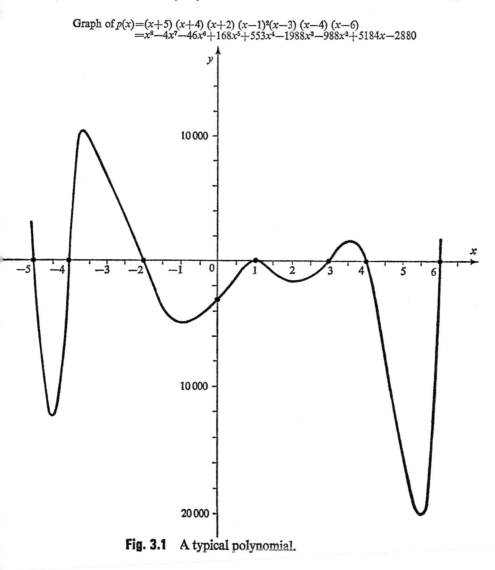

Fig. 3.1 A typical polynomial.

loss of speed in the final stage of the iteration. The method of false position is a simple method suitable for finding a real root and Bairstow's method is frequently used to find a conjugate-complex pair. The theoretical justification for Bairstow's method is given in Balfour and McTernan (1967) and Bull (1966), and can easily be understood by those familiar with partial differentiation. Only the computational steps in the process are given here.

3.4.2 Bairstow's method
Initially, an approximation $x^2+p_0x+q_0$ to a quadratic factor is required. The process is one of iteration to find increments Δp_r and Δq_r which improve the approximation. The sequence of values

$$p_{r+1}=p_r+\Delta p_r, \qquad q_{r+1}=q_r+\Delta q_r, \qquad r=0,1,2,\ldots, \qquad (3.26)$$

are found by solving two simultaneous equations

$$a_{11}\Delta p_r + a_{12}\Delta q_r = b_1$$
$$a_{21}\Delta p_r + a_{22}\Delta q_r = b_2 \tag{3.27}$$

The steps in the process are:

1 The polynomial $f(x)$ is divided by the current approximation to the quadratic $x^2 + p_r x + q_r$ to give a quotient $Q(x)$ and remainder $R_1 x + S_1$. When the process has converged both R_1 and S_1 will be zero.

2 The function $xQ(x)$ is then divided by $x^2 + p_r x + q_r$ to give a remainder $R_2 x + S_2$.

3 The function $Q(x)$ is divided by $x^2 + p_r x + q_r$ to give a remainder $R_3 x + S_3$. It should be noted that this does not involve further arithmetic since the numbers are virtually the same as in step 2 but in different columns.

4 The coefficients in the two simultaneous equations are given by $a_{11} = -R_2$, $a_{12} = -R_3$, $a_{21} = -S_2$, $a_{22} = -S_3$, and $b_1 = -R_1$ and $b_2 = -S_1$.

5 The new values of p_{r+1} and q_{r+1} are found and the process repeated until the values of p and q have converged. One feature of the scheme which has been ignored, since there is no simple solution, is the problem of finding the initial approximation to the quadratic factor. One possibility would be to use the factor corresponding to the three terms of lowest degree in the equation, although there is no guarantee convergence will result.

Example 3.5 gives a numerical table showing the use of Bairstow's method.

Another method which has attracted favourable attention is Laguerre's method which is defined by the following formulae.

$$x_{n+1} = x_n - \frac{nf(x_n)}{f'(x_n) \pm [H(x_n)]^{\frac{1}{2}}} \tag{3.28}$$
$$H(x_n) = (n-1)^2[f'(x_n)]^2 - n(n-1)f(x_n)f''(x_n)$$

This method gives two sets of iterates which converge to the nearest zero greater than a and the nearest zero less than a where a is the first approximation. The method has convergence of order 3, if the sign of the square root is chosen to agree with that of $f'(x_n)$, but this is obtained at the expense of some extra computation. The attraction of the method is because some workers suggest that this method is more likely to converge than Newton's method when the initial approximation is not in the immediate neighbourhood of the root. The method is more feasible with a polynomial than for a general $f(x)$ since the second derivative $f''(x_n)$ can be evaluated by three applications of the synthetic-division process, which is a simple efficient computer algorithm.

3.5 Methods for finding all roots

The last section pointed out the problems of error build-up which occur when the process of iteration, followed by deflation, is used repetitively. It would, therefore, be attractive to find all the roots of a polynomial simultaneously. Unfortunately, methods of this kind have the disadvantage that, when the results are in the neighbourhood of the correct values, the convergence rate is usually much slower than that achieved by a good single-root method. A good strategy is to use a method which finds all roots to give good initial approximations, combined with a strongly convergent method to improve the accuracy of individual roots.

3.5.1 Quotient difference algorithm

This method was first proposed by Rutishauser (1956) and is described simply in the books of Henrici (1964), and Balfour and McTernan (1967). The theoretical background requires a knowledge of complex-function theory and is, therefore, beyond the scope of this book. However, in view of the importance of the method and the simple nature of the calculations the computational steps are presented here.

For a polynomial of degree n, a table is drawn up with $2n + 1$ columns. The two outermost columns of the table are zero. The remaining columns are in two sets of alternate columns which are obtained by two separate calculation rules, a quotient rule and a difference rule. The method of generating the table is most easily explained by building up from the left-most column of zeros using a scheme based on Bernouilli's method of finding roots. (This is not discussed in this text.) However, in order to minimize the build-up of errors, it is preferable to generate the table, row by row. This requires some method of generating the first two rows which will appear rather arbitrary to the reader. The description by Henrici (1964) will provide more background to these rules if required.

The table for a cubic is shown in Table 3.2.

Table 3.2

The quotient–difference table

$q_0^{(1)}$		$q_{-1}^{(2)}$		$q_{-2}^{(3)}$	
0	$\varepsilon_0^{(1)}$		$\varepsilon_{-1}^{(2)}$		0
$q_1^{(1)}$		$q_0^{(2)}$		$q_{-1}^{(3)}$	
0	$\varepsilon_1^{(1)}$		$\varepsilon_0^{(2)}$		0
$q_2^{(1)}$		$q_1^{(2)}$		$q_0^{(3)}$	
0	$\varepsilon_2^{(1)}$		$\varepsilon_1^{(2)}$		0
$q_3^{(1)}$		$q_2^{(2)}$		$q_1^{(3)}$	
$\vdots$		$\vdots$		$\vdots$	

Note that the superscripts in each column are constant and the subscripts are constant along the forward diagonal. For the general scheme of degree n the process is started by calculating the elements of the first two rows as follows:

$$q_0^{(1)} = -a_{n-1}/a_n$$
$$q_{1-r}^{(r)} = 0, \qquad r = 2, 3, \ldots, n \qquad (3.29)$$
$$\varepsilon_{1-r}^{(r)} = a_{r-1}/a_r, \qquad r = n-1, n-2, \ldots, 1$$

The elements are then calculated moving right along the rows and down the columns using the four values at the points of a rhombus as follows:

$$\alpha_j$$
$$\beta_{j+1} \qquad\qquad\qquad \beta_j$$
$$\alpha_{j+1}$$

If the α_j elements lie in a q column then $\alpha_j + \beta_j = \beta_{j+1} + \alpha_{j+1}$. If the α_j elements lie in an ε column then $\alpha_j \times \beta_j = \beta_{j+1} \times \alpha_{j+1}$. Solving row by row, all the quantities α_j, β_j, and β_{j+1}, are known and the unknown α_{j+1} is found from one of the two equations

$$q_{j+1}^{(r)} = q_j^{(r)} + \varepsilon_j^{(r)} - \varepsilon_{j+1}^{(r-1)} \tag{3.30}$$

$$\varepsilon_{j+1}^{(r)} = \varepsilon_j^{(r)} \cdot \frac{q_j^{(r+1)}}{q_{j+1}^{(r)}} \tag{3.31}$$

It should be noted that the scheme may break down completely if any $q_j^{(r)}$ becomes equal to zero since the division cannot be carried out. The above algorithm is most useful when the roots of the polynomial z_r $(r = 1, 2, \ldots, n)$ satisfy the relation $|z_1| > |z_2| > \cdots > |z_n|$. In this case it can be proved that

1 $\displaystyle\lim_{j \to \infty} q_j^{(r)} = z_r, \qquad r = 1, 2, \ldots, n \tag{3.32a}$

2 $\displaystyle\lim_{j \to \infty} \varepsilon_j^{(r)} = 0, \qquad r = 1, 2, \ldots, n \tag{3.32b}$

Thus, the scheme will reveal if the roots are all of unequal modulus by studying the behaviour of the ε columns.

If some roots have equal modulus, i.e., $|z_1| \geqslant |z_2| \geqslant \cdots \geqslant |z_n|$ then the following conditions hold:

1 For every r such that $|z_{r-1}| > |z_r| > |z_{r+1}|$

$$\lim_{j \to \infty} q_j^{(r)} = z_r \tag{3.33a}$$

2 For every r such that $|z_r| > |z_{r+1}|$

$$\lim_{j \to \infty} \varepsilon_j^{(r)} = 0 \tag{3.33b}$$

Thus, the table is divided into groups of q columns, corresponding to roots of equal modulus, by the ε columns which tend to zero.

The most common case of roots of equal modulus occurs when an equation with real coefficients has a conjugate-complex pair. There will be two q columns separated by an ε column which does not tend to zero and the two roots required, z_{r+1} and z_{r+2}, are the solutions of the quadratic equation

$$z^2 - A_r z + B_r = 0$$

where

$$\lim_{j \to \infty} (q_{j+1}^{(r+1)} + q_j^{(r+2)}) = A_r$$
$$\lim_{j \to \infty} (q_j^{(r+1)} q_j^{(r+2)}) = B_r \tag{3.34}$$

The reader interested in more complex examples should refer to the quoted texts. Numerical results are given in Example 3.6.

3.5.2 Lehmer–Schur method

Although this method is always convergent it is extremely laborious and very slow to converge. It is, of course, suitable for finding first approximations. The method is based on a sequence of calculations which test whether a root of the equation lies within a given test circle. A sequence of test circles must be produced which can be used to search the whole complex plane, and which will converge on the root when this has been isolated.

Firstly, the test is used to determine whether a root exists within a circle centred at the origin with unit radius. If this test is unsuccessful a transformation of the original equation will enable a search to be made in a circle of twice the radius. If still unsuccessful a further transformation gives a search of a circle four times the original radius etc. This process is repeated until a root is found which lies within an annulus $R < |z| < 2R$. The area in which the root lies is now narrowed by searching in eight overlapping circles of radius $4R/5$ with centres at the points

$$z = \frac{3R\,e^{2\pi i k/8}}{2\cos(\pi/8)}, \qquad k = 0, 1, \ldots, 7 \qquad (3.35)$$

One of these circles must contain a root, although it is important to realize that more than one of the circles may contain a root, since the overlapping circles cover a wider area than the annulus and also cover some areas twice. We now have a circle which contains a root, so a search in succession of circles, each having a radius half the previous one, will yield an annulus which contains a root and then the process can once again be repeated.

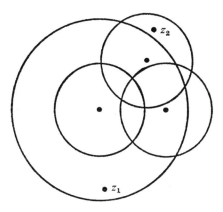

Fig. 3.2   The Lehmer–Schur method.

The mathematical details and the complete algorithm for this method are available in Ralston (1965), or in the original paper by Lehmer (1961), but the reader is advised to use a standard library program on a computer if this is available. The main difficulties with the method arise when several roots lie in a search area or when the search isolates a root different from that which determined the annulus. This can easily be seen by reference to Fig. 3.2 which shows that the method having found a root z_1 in the annulus would eventually converge to a different root, z_2.

3.5.3   Graeffe's root squaring method

In its simplest form this method takes an equation whose roots are unequal in magnitude and from this derives a second equation with roots which are the square of the original roots. The process is repeated until the new roots are

2^m times the original roots. For sufficiently large m the magnitude of the original roots can be determined by a simple calculation.

The method is attractive because the squaring process increases the separation of the roots which simplifies the root finding process. Error analysis shows that it is difficult to obtain accurate values for the roots when they are close together. Although in its simplest form the method cannot deal with the case where roots have equal magnitude there exist modifications to the method which enables this to be achieved.

Let the coefficients of the various polynomials which are obtained in Graeffe's method be denoted by $a_r^{(j)}$ where $a_r^{(0)} = a_r$ as defined in Eqn (3.1). The coefficients of the new polynomial are obtained as follows.

$$
\begin{aligned}
a_n^{(1)} &= [a_n^{(0)}]^2 \\
a_{n-1}^{(1)} &= -[a_{n-1}^{(0)}]^2 + 2a_n^{(0)}a_{n-2}^{(0)} \\
a_{n-2}^{(1)} &= [a_{n-2}^{(0)}]^2 - 2a_{n-1}^{(0)}a_{n-3}^{(0)} + 2a_n^{(0)}a_{n-4}^{(0)}
\end{aligned} \tag{3.36}
$$

and a similar rule with different superscripts will generate the further members of the sequence. There is a standard relationship between the coefficients of a polynomial and the sums of the products of the roots Z_r. For example,

$$
\begin{aligned}
\frac{a_{n-1}}{a_n} &= -\sum_{r=1}^{n} Z_r \\
\frac{a_{n-2}}{a_n} &= \sum_{r=1}^{n} \sum_{s=r+1}^{n} Z_r Z_s \\
\frac{a_0}{a_n} &= (-1)^n Z_0 . Z_1 \cdots Z_n
\end{aligned} \tag{3.37}
$$

where the summation is taken over all possible combinations of the values Z_r, Z_s between 1 and n. In the case under discussion, due to the power operation on the roots, many of the terms become negligible. Therefore, the following approximations can be used for sufficiently high polynomials in the Graeffe sequence. For simplicity we assume $|Z_1| > |Z_2| > \cdots > |Z_n|$

$$
\begin{aligned}
Z_1^{2j} &\approx -a_{n-1}/a_n \\
Z_1^{2j}Z_2^{2j} &\approx a_{n-2}/a_n \\
&\cdots\cdots\cdots\cdots\cdots\cdots\cdots\cdots\cdots \\
Z_1^{2j}Z_2^{2j} \cdots Z_n^{2j} &\approx (-1)^n(a_0/a_n)
\end{aligned} \tag{3.38}
$$

and, by taking the appropriate root, the values of the various roots Z_r can be found. The sign of the Z_r is found by substituting in the original equation. The reader should not be misled by the over-simplified explanation of the Graeffe process considered above. The problems caused by roots of equal magnitude and poor convergence in the process must be considered before an adequate computer program can be produced.

The interested reader is strongly recommended to read the chapter by Bareiss, in Ralston and Wilf, Volume 2 (1967) which gives a great deal of valuable discussion on the problem of solving polynomials and gives a detailed flow chart for using the root-squaring process in conjunction with a further procedure. The reader is not advised to program this method unless he has a firm grasp of the computer problems which must be overcome.

3.6 Comparison of methods

The most satisfactory general-purpose program would appear to be that described by Bareiss (1967). This method shows much faster convergence than the Lehmer–Schur method although the arithmetic involved is more complicated. The simplest approach for anyone writing their own program would be to use the quotient–difference algorithm to obtain approximations to the roots, and methods with good convergence properties to improve the root, such as Newton's for single roots and Bairstow's for complex-conjugate roots. If a standard library program were available for the Lehmer–Schur method this could be used instead of the quotient difference algorithm. The complete solution of a polynomial is not trivial and where a reliable computer program is available this should be used. However, the user should realize that there will still be polynomials which cannot be solved numerically by standard means and require careful mathematical analysis to obtain meaningful answers.

Bibliographic notes

The solution of polynomials is covered in some detail in Ralston (1965), Wilkinson (1963), and Bareiss (1967). The latter text gives both the mathematical background and the computational detail including flow charts. Numerical examples of the synthetic division process and other polynomial arithmetic are given in Bull (1966) and Redish (1961). A wide coverage at a level suitable for the non-mathematician is given in Balfour and McTernan (1967). The quotient difference algorithm is described in some detail in Henrici (1964). The material is presented formally and includes several theorems on the various properties of the method. A simpler treatment is given in Balfour and McTernan (1967).

Worked examples

1 Find the value of the polynomial $5x^4 - 2x^3 + x - 1$ and its derivative at the point $x = 2$ by using the remainder theorem and the synthetic-division process.
    The remainder after division by $x - 2$ gives the value at the point $x = 2$.
The remainder, after dividing the quotient by $x - 2$, gives the value of the derivative.

$$
\begin{array}{r|rrrrr}
 & 5 & -2 & 0 & 1 & -1 \\
2 & 0 & 10 & 16 & 32 & 66 \\
\hline
 & 5 & 8 & 16 & 33 & 65 = f(2) \\
2 & 0 & 10 & 36 & 104 & \\
\hline
 & 5 & 18 & 52 & 137 = f'(2) &
\end{array}
$$

2 Two of the roots of the equation $x^4 + 2x^3 - 6x^2 - 22x - 15$ are the conjugate-complex pair $-2 + i, -2 - i$ which correspond to a quadratic factor $x^2 + 4x + 5$. Synthetic division by the quadratic factor would deflate the polynomial ready for a further iterative solution.

$$
\begin{array}{r|rrrrr}
 & 1 & 2 & -6 & -22 & -15 \\
-4 & 0 & -4 & 8 & 12 & 0 \\
-5 & 0 & 0 & -5 & 10 & +15 \\
\hline
 & 1 & -2 & -3 & 0 & 0
\end{array}
$$

The remainders confirm that $x^2 + 4x + 5$ is a factor of the polynomial, the other factor being $x^2 - 2x - 3$.

3 Use the Sturm sequence technique to find the approximate location of one of the roots of $x^3 - 3x^2 + 2x - 1$.

$$f_0(x) = x^3 - 3x^2 + 2x - 1$$
$$f_1(x) = 3x^2 - 6x + 2$$
$$f_2(x) = 2x/3 + \tfrac{1}{3}$$
$$f_3(x) = -\tfrac{23}{4}$$

The working for the last two entries is given by the following two synthetic-division tables

$$
\begin{array}{r|rrrr}
 & 1 & -3 & 2 & -1 \\
2 & 0 & 2 & -2 & 0 \\
-\tfrac{2}{3} & 0 & 0 & -\tfrac{2}{3} & +\tfrac{2}{3} \\
\hline
 & 1 & -1 & -\tfrac{2}{3} & -\tfrac{1}{3}
\end{array}
$$

$$
\begin{array}{r|rrr}
 & 3 & -6 & 2 \\
-\tfrac{1}{2} & 0 & -\tfrac{3}{2} & \tfrac{15}{4} \\
\hline
 & 3 & -\tfrac{15}{2} & \tfrac{23}{4}
\end{array}
$$

The signs of these functions are plotted at various points to give the following table.

x	$-\infty$	0	∞	1	2	3
$f_0(x)$	$-$	$-$	$+$	$-$	$-$	$+$
$f_1(x)$	$+$	$+$	$+$	$+$	$+$	$+$
$f_2(x)$	$-$	$+$	$+$	$+$	$+$	$+$
$f_3(x)$	$-$	$-$	$-$	$-$	$-$	$-$
	2	2	1	2	2	1

Hence, the root lies between $x = 2$ and $x = 3$.

4 Use the Sturm sequence technique to find the approximate location of the roots of the polynomial $x^4 - 9x^2 + 4x + 12$. The Sturm table is developed as follows.

$$f_0(x) = x^4 - 9x^2 + 4x + 12$$
$$f_1(x) = 4x^3 - 18x + 4$$
$$f_2(x) = 18x^2/4 - 3x - 12$$
$$f_3(x) = 50x/9 - \tfrac{100}{9}$$

When this factor is divided into $f_2(x)$ there is no remainder, which indicates a double root, given by the factor $f_3(x)$, i.e., $x - 2$. The modified Sturm sequence is obtained by dividing through by this factor.

$$f_0^*(x) = x^3 + 2x^2 - 5x - 6$$
$$f_1^*(x) = 4x^2 + 8x - 2$$
$$f_2^*(x) = 18x/4 + 6$$
$$f_3^*(x) = \tfrac{50}{9}$$

The following table shows the position of the roots.

x	$-\infty$	-4	-2	0	∞
$f_0^*(x)$	$-$	$-$	$+$	$-$	$+$
$f_1^*(x)$	$+$	$+$	$-$	$-$	$+$
$f_2^*(x)$	$-$	$-$	$-$	$+$	$+$
$f_3^*(x)$	$+$	$+$	$+$	$+$	$+$
	3	3	2	1	0

There is one root between -4 and -2, one root between -2 and 0, and one root between 0 and ∞. This latter root is the double root at $x = 2$ which has already been found. Thus, the approximate position of all the roots is now established.

5 Use Bairstow's method to find a quadratic factor of the polynomial.

$$P(x) = 2x^6 + 15x^5 + 43x^4 + 51x^3 + 41x^2 + 108x + 160$$

The last three coefficients are used to suggest a first approximation $x^2 + 3x + 4$. In Bairstow's method we use synthetic division to divide $P(x)$ by the trial factor, then $Q(x)$ and $xQ(x)$.

$$P(x) = Q(x)(x^2 + 3x + 4) + R_1 x + S_1$$
$$Q(x) = B(x)(x^2 + 3x + 4) + R_3 x + S_3$$
$$xQ(x) = C(x)(x^2 + 3x + 4) + R_2 x + S_2$$

The increments to the coefficients are found from the equations

$$-R_2 \Delta p_0 - R_3 \Delta q_0 = -R_1$$
$$-S_2 \Delta p_0 - S_3 \Delta q_0 = -S_1$$

and the next approximations are $3 + \Delta p_1$ and $4 + \Delta q_1$. For convenience all the negative signs can be removed. The above process is then repeated with the new quadratic factor.

	2	15	43	51	41	108	160
-3	0	-6	-27	-24	27	-108	0
-4	0	0	-8	-36	-32	36	-144
	2	9	8	-9	36	36	16

Therefore,

$$Q(x) = 2x^4 + 9x^3 + 8x^2 - 9x + 36, \qquad R_1 = 36, \qquad S_1 = 16$$

	2	9	8	-9	36	0
-3	0	-6	-9	27	-18	0
-4	0	0	-8	-12	36	-24
$Q(x)$	2	3	-9	$\mid$ 6	72	
$xQ(x)$				6 $\mid$	54	-24

$$R_2 = 54, \qquad S_2 = -24, \qquad R_3 = 6, \qquad S_3 = 72$$

Therefore, the two simultaneous equations become

$$54 \Delta p_0 + 6 \Delta q_0 = 36$$
$$-24 \Delta p_0 + 72 \Delta q_0 = 16$$
$$\Delta p_0 = 0{\cdot}6, \qquad \Delta q_0 = 0{\cdot}4$$

Note, that it is unnecessary to use high accuracy in the initial stages of iteration when the values are still inaccurate. The number of figures used in the calculation will be increased as closer approximations are obtained. The new approximation to the quadratic factor is $x^2 + 3{\cdot}6x + 4{\cdot}4$ and the above sequence of the calculations are repeated.

	2	15	43	51	41	108	160
$-3{\cdot}6$	0	$-7{\cdot}2$	$-28{\cdot}08$	$-22{\cdot}03$	$19{\cdot}26$	$-120{\cdot}02$	0
$-4{\cdot}4$	0	0	$-8{\cdot}80$	$-34{\cdot}32$	$-26{\cdot}92$	$23{\cdot}53$	$-146{\cdot}69$
	2	$7{\cdot}8$	$6{\cdot}12$	$-5{\cdot}35$	$33{\cdot}34$	$\mid$ $11{\cdot}51$	$13{\cdot}31$

$$Q(x) = 2x^4 + 7{\cdot}8x^3 + 6{\cdot}12x^2 - 5{\cdot}35x + 33{\cdot}34, \qquad R_1 = 11{\cdot}51, \qquad S_1 = 13{\cdot}31$$

	2	$7{\cdot}80$	$6{\cdot}12$	$-5{\cdot}35$	$33{\cdot}34$	$0{\cdot}00$
$-3{\cdot}6$	$0{\cdot}00$	$-7{\cdot}20$	$-2{\cdot}16$	$17{\cdot}42$	$-33{\cdot}94$	$0{\cdot}00$
$-4{\cdot}4$	$0{\cdot}00$	$0{\cdot}00$	$-8{\cdot}80$	$-2{\cdot}64$	$21{\cdot}29$	$-41{\cdot}49$
	2	$0{\cdot}60$	$-4{\cdot}84$	$\mid$ $9{\cdot}43$	$54{\cdot}63$	
				$9{\cdot}43$ $\mid$	$20{\cdot}69$	$-41{\cdot}49$

$$R_2 = 20{\cdot}69, \qquad S_2 = -41{\cdot}49, \qquad R_3 = 9{\cdot}43, \qquad S_3 = 54{\cdot}63$$

The two simultaneous equations are then solved.

$$20 \cdot 69 \, \Delta p_1 + 9 \cdot 43 \, \Delta q_1 = 11 \cdot 51$$
$$-41 \cdot 49 \, \Delta p_1 + 54 \cdot 63 \, \Delta q_1 = 13 \cdot 31$$
$$\Delta p_1 = 0 \cdot 33, \qquad \Delta q_1 = 0 \cdot 49$$

The new quadratic factor is $x^2 + 3 \cdot 93x + 4 \cdot 89$ leading to the simultaneous equations

$$15 \cdot 981 \, \Delta p_2 + 6 \cdot 366 \, \Delta q_2 = 1 \cdot 782$$
$$-31 \cdot 129 \, \Delta p_2 + 40 \cdot 999 \, \Delta q_2 = 2 \cdot 337$$

with solutions $\Delta p_2 = 0 \cdot 068$ and $\Delta q_2 = 0 \cdot 108$.
The new factor $x^2 + 3 \cdot 998x + 4 \cdot 998$ leads to the equations

$$16 \cdot 8943 \, \Delta p_3 + 5 \cdot 0654 \, \Delta q_3 = 0 \cdot 0438$$
$$-25 \cdot 3168 \, \Delta p_3 + 37 \cdot 1457 \, \Delta q_3 = -0 \cdot 0226$$

with solutions $\Delta p_3 = 0 \cdot 0020$ and $\Delta q_3 = 0 \cdot 0019$, and a new factor
$$x^2 + 4 \cdot 0000x + 4 \cdot 9999.$$
The simultaneous equations are

$$17 \cdot 002 \, 81 \, \Delta p_4 + 4 \cdot 999 \, 00 \, \Delta q_4 = 0 \cdot 000 \, 45$$
$$-24 \cdot 994 \, 50 \, \Delta p_4 + 36 \cdot 998 \, 81 \, \Delta q_4 = 0 \cdot 003 \, 65$$

giving $\Delta p_4 = 0 \cdot 000 \, 00$ and $\Delta q_4 = 0 \cdot 000 \, 09$, and a factor $x^2 + 4 \cdot 000 \, 00x + 4 \cdot 999 \, 99$.

Division by this quadratic factor gives a remainder of $0 \cdot 000 \, 00x + 0 \cdot 000 \, 32$ and thus, the process has virtually converged to the correct factor. After division by this factor a quartic equation results and further use of Bairstow's method will produce the remaining two factors.

6 The table on page 55 shows the results obtained by applying the quotient difference algorithm to the polynomial.

$$P(x) = (x + 2)(x + 1)(x - 1)(x - 2)(x - 4) = x^5 - 4x^4 - 5x^3 + 20x^2 + 4x - 16$$

The root $x = 4 \cdot 0$ is clearly shown between two columns of zeros. The two pairs of roots of equal modulus generate five columns of which the two outside columns are zero and columns two and four give the values required to calculate the roots. We have

$$A = \lim_{j \to \infty} (q_{j+1}^{(r+1)} + q_j^{(r+2)}) = 0$$
$$B = \lim_{j \to \infty} (q_j^{(r+1)} q_j^{(r+2)}) = -4$$

The roots are the solutions of the equation

$$x^2 - 4 = 0 \quad \text{which gives} \quad x = \pm 2$$

Similarly, the second pair have

$$A = 0, \qquad B = -1$$

giving $x = \pm 1$.

PROGRAM ENTERED CLOCKED: MINS 00:27 SECS

0·00000	4·00000	0·00000	0·00000	-4·00000	0·00000	0·20000	0·00000	-4·00000	0·00000	0·00000
0·00000	5·25000	1·25000	-5·25000	3·20000	4·20000	-0·20000	-4·20000	3·80952	4·00000	0·00000
0·00000	4·00000	-1·25000	-0·80000	-3·20000	0·80000	0·04762	-0·19048	-3·80952	0·19048	0·00000
0·00000	4·25000	0·25000	-4·25000	3·04762	4·04762	-0·04762	-4·04762	3·76471	4·00000	0·00000
0·00000	4·00000	-0·25000	-0·95238	-3·04762	0·95238	0·01176	-0·23529	-3·76471	0·23529	0·00000
0·00000	4·05952	0·05952	-4·05952	3·01176	4·01176	-0·01176	-4·01176	3·75367	4·00000	0·00000
0·00000	4·00000	-0·05952	-0·98824	-3·01176	0·98824	0·00293	-0·24633	-3·75367	0·24633	0·00000
0·00000	4·01471	0·01471	-4·01471	3·00293	4·00293	-0·00293	-4·00293	3·75092	4·00000	0·00000
0·00000	4·00000	-0·01471	-0·99707	-3·00293	0·99707	0·00073	-0·24908	-3·75092	0·24908	0·00000
0·00000	4·00367	0·00367	-4·00367	3·00073	4·00073	-0·00073	-4·00073	3·75023	4·00000	0·00000
0·00000	4·00000	-0·00367	-0·99927	-3·00073	0·99927	0·00018	-0·24977	-3·75023	0·24977	0·00000
0·00000	4·00092	0·00092	-4·00092	3·00018	4·00018	-0·00018	-4·00018	3·75006	4·00000	0·00000
0·00000	4·00000	-0·00092	-0·99982	-3·00018	0·99982	0·00005	-0·24994	-3·75006	0·24994	0·00000
0·00000	4·00023	0·00023	-4·00023	3·00005	4·00005	-0·00005	-4·00005	3·75001	4·00000	0·00000
0·00000	4·00000	-0·00023	-0·99995	-3·00005	0·99995	0·00001	-0·24999	-3·75001	0·24999	0·00000
0·00000	4·00006	0·00006	-4·00006	3·00001	4·00001	-0·00001	-4·00001	3·75000	4·00000	0·00000
0·00000	4·00000	-0·00006	-0·99999	-3·00001	0·99999	0·00000	-0·25000	-3·75000	0·25000	0·00000
0·00000	4·00001	0·00001	-4·00001	3·00000	4·00000	-0·00000	-4·00000	3·75000	4·00000	0·00000
0·00000	4·00000	-0·00001	-1·00000	-3·00000	1·00000	0·00000	-0·25000	-3·75000	0·25000	0·00000
0·00000	4·00000	0·00000	-4·00000	3·00000	4·00000	-0·00000	-4·00000	3·75000	4·00000	0·00000
0·00000	4·00000	-0·00000	-1·00000	-3·00000	1·00000	0·00000	-0·25000	-3·75000	0·25000	0·00000
0·00000	4·00000	0·00000	-4·00000	3·00000	4·00000	0·00000	-4·00000	3·75000	4·00000	0·00000

Problems

1 Evaluate the polynomial $4x^4 - 2x^3 + 5x - 3$ at the point $x = -2$ by the nested-multiplication method.

2 Use a Sturm sequence to find the approximate location of one root of
$$x^3 - 3x^2 + x - 1.$$

3 Use Bairstow's method to find the two quadratic factors of
$$2x^4 + 7x^3 + 5x^2 - 4x + 32.$$
As a first approximation use the last three terms of the equation.

4

Solution of linear simultaneous equations

4.1 Introduction

There are many physical and numerical problems in which the solution is obtained by solving a set of linear simultaneous equations. This problem can be a fairly simple one, when the number of unknowns is small, and is often studied at an elementary level in mathematics. However, there are problems when the number of independent equations is different from the number of unknowns. The problem has a unique solution when there are n linearly independent equations and n unknowns. When there are less than n such equations then a unique solution cannot be obtained.

If there are more than n equations then two possibilities exist. The extra equations may arise because some equations are linearly dependent on the others. In this case the problem can be solved by selecting the n equations which are independent. The alternative is that some of the equations are not consistent with each other. This often arises when attempting to fit some chosen mathematical form to a set of experimental results. In an attempt to cater for experimental error a much larger number of observations is taken than is necessary to determine uniquely the parameters in the equation. The technique to deal with this situation is the method of least squares which is covered in Chapter 7.

In this chapter attention is restricted to the case where we have n linear simultaneous equations in n unknowns which we will write in the form

$$
\begin{aligned}
a_{11}x_1 + a_{12}x_2 + \cdots + a_{1n}x_n &= b_1 \\
a_{21}x_1 + a_{22}x_2 + \cdots + a_{2n}x_n &= b_2 \\
&\cdots\cdots\cdots\cdots\cdots\cdots\cdots \\
a_{n1}x_1 + a_{n2}x_2 + \cdots + a_{nn}x_n &= b_n
\end{aligned}
\tag{4.1}
$$

The double-suffix notation is quite simple if one remembers that the first suffix specifies which row the element appears in and the second suffix specifies the number of the column.

Many readers will be familiar with matrix notation in which a set of coefficients such as the a_{ij} $(i, j = 1, 2, \ldots, n)$ are represented by a single quantity $\mathbf{A}$ and the elements x_j $(j = 1, 2, \ldots, n)$ are represented by a vector $\mathbf{X}$. Using the rules of matrix algebra the set of equations can be represented in the form

$$
\mathbf{AX} = \mathbf{B}
\tag{4.2}
$$

Mathematically, these equations have a unique solution if $\det(\mathbf{A})$ is not equal to zero. Numerically, there is a difficulty since the concept of zero is

C

rather imprecise in computer terms. The result of a calculation to which the answer should be zero may well turn out to be 10^{-50}, for example, as a result of data or round-off error. It is clear that a very small value for the determinant is likely to be as troublesome as a zero value, since the two may well be indistinguishable.

Such sets of equations are called ill-conditioned; they are characterized by the fact that a small change in the constants in the equation can cause a large change in the answer. As an example, the following sets of equations give approximately 200% variation in the answer for approximately 1% variation in

$$
\begin{array}{lll}
x + 100y = 3 & x + 101y = 3 & x + 100y = 3 \\
x + 100y = 6 & x + 100y = 6 & x + 101y = 6 \\
\hline
\text{No solution} & y = -3 & y = 3 \\
& x = 306 & x = -297
\end{array}
\tag{4.3}
$$

the coefficients. Graphically, we can see that the lines are nearly parallel so that a very small variation in slope causes a marked difference in the point of intersection.

There are two types of method which will be discussed for the solution of Eqns (4.1). The direct methods are based on elimination techniques very similar to those used at school level for the solution of a few equations. The indirect or iterative methods are based on calculating a sequence of approximations which hopefully converge to a value sufficiently close to the true solution. It is important to the user that he appreciates the properties of these two types of method so that an intelligent choice of method can be used in a particular situation. The advantage of direct methods is that the amount of computation is fixed and can be determined beforehand, but in the case of iterative methods the calculation must continue indefinitely until the answers have converged to sufficient accuracy. Indeed, it is quite possible that iterative methods may not converge at all.

A further factor which must be considered is the number of coefficients which have the value zero. For example, in many problems which arise in the solution of ordinary and partial-differential equations, a large set of simultaneous equations is produced each of which has only a few nonzero coefficients. Such a matrix is called a sparse matrix. It will be seen that this increases the attractiveness of iterative methods considerably, since the work involved is directly proportional to the number of nonzero elements.

A word of caution must however be issued in the case of some sparse equations which have certain very simple structures. In these cases it is possible to design a direct method in which the amount of computation involved is directly proportional to the number of nonzero elements. This advantage of iterative methods is then lost and the finite nature of the direct method makes this more suitable.

4.2 Direct methods

4.2.1 Gaussian elimination

The basis of the direct method is the simple elimination of unknowns, familiar to many from practice in school examples. In a typical hand calculation this is done in a haphazard fashion using the skill of the man concerned to find the simplest order of elimination of the unknowns. There are two disadvantages here; namely, the haphazard ordering of operations is unsuitable for calculation on a computer and, a systematic ordering of the calculations is necessary so that if a breakdown in the calculations takes place it is possible to identify the cause of the problem. It is a familiar situation where considerable hand calculation reveals the answer $0 = 0$ and there is little indication as to whether the original problem was incorrectly formulated, or whether the steps of the calculation were performed incorrectly.

The simple scheme known as gaussian elimination will be described first, but the reader should bear in mind that for a scheme to be computationally sound, it should include routines both for scaling the matrix and for partial pivoting during the elimination. These refinements are discussed later in this section. In this simple treatment we assume that all the divisors in the scheme are nonzero. The equations to be solved are

$$a_{11}x_1 + a_{12}x_2 + \cdots + a_{1n}x_n = b_1$$
$$a_{21}x_1 + a_{22}x_2 + \cdots + a_{2n}x_n = b_2$$
$$\cdots\cdots\cdots\cdots\cdots\cdots\cdots\cdots \tag{4.4}$$
$$a_{n1}x_1 + a_{n2}x_2 + \cdots + a_{nn}x_n = b_n$$

The first equation is stored for later use and the variable x_1 is eliminated from the remaining $n-1$ equations by subtracting an appropriate multiple of the first equation from each of the other equations. Let the original coefficients be given the notation

$$a_{ij}^{(1)} = a_{ij}, \qquad i, j = 1, 2, \ldots, n \tag{4.5}$$
$$b_i^{(1)} = b_i, \qquad i = 1, 2, \ldots, n \tag{4.6}$$

The new coefficients are found by using the multipliers

$$m_{i1} = a_{i1}^{(1)}/a_{11}^{(1)}, \qquad i = 2, 3, \ldots, n \tag{4.7}$$

and forming the new elements

$$a_{ij}^{(2)} = a_{ij}^{(1)} - m_{i1}a_{1j}^{(1)}, \qquad i = 2, 3, \ldots, n$$
$$j = 1, 2, \ldots, n \tag{4.8}$$
$$b_i^{(2)} = b_i^{(1)} - m_{i1}b_1^{(1)}, \qquad i = 2, 3, \ldots, n \tag{4.9}$$

It can be seen that the elements in the first column, $j = 1$, have the values

$$a_{i1}^{(2)} = a_{i1}^{(1)} - \frac{a_{i1}^{(1)}}{a_{11}^{(1)}} \cdot a_{11}^{(1)} = 0 \tag{4.10}$$

so that the first variable x_1 has been eliminated in the last $n-1$ equations. If the first row is now ignored the equations have the same form as Eqns (4.4) but with one row and one column less.

If the previous procedure is repeated until it has been performed $n-1$ times altogether the remaining equation will have only one unknown and can be solved very easily. At each stage in the process when the variable x_k is to be eliminated the multipliers are formed

$$m_{ik} = a_{ik}^{(k)}/a_{kk}^{(k)}, \qquad i = k+1, k+2, \ldots, n \tag{4.11}$$

and new elements formed,

$$a_{ij}^{(k+1)} = a_{ij}^{(k)} - m_{ik}a_{kj}^{(k)} \qquad i = k+1, k+2, \ldots, n \qquad (4.12)$$
$$j = k, k+1, \ldots, n$$
$$b_i^{(k+1)} = b_i^{(k)} - m_{ik}b_k^{(k)} \qquad i = k+1, k+2, \ldots, n \qquad (4.13)$$

The result of this elimination process is an upper-triangular set of equations given by

$$
\begin{aligned}
a_{11}^{(1)}x_1 + a_{12}^{(1)}x_2 + \cdots + a_{1n}^{(1)}x_n &= b_1^{(1)} \\
a_{22}^{(2)}x_2 + \cdots + a_{2n}^{(2)}x_n &= b_2^{(2)} \\
\cdots\cdots\cdots\cdots\cdots \\
a_{nn}^{(n)}x_n &= b_n^{(n)}
\end{aligned}
\qquad (4.14)
$$

where all the elements below the diagonal are zero. It is easy to solve these equations by a process of back substitution. The last equation has the solution

$$x_n = b_n^{(n)}/a_{nn}^{(n)}$$

and this value can then be substituted in the next lowest equation to give x_{n-1} etc.

By working back up the equations the values of all the variables can be calculated. A table of results for a gaussian elimination is given in Example 4.1.

4.2.2 Multiple right-hand sides

It frequently happens that several equations must be solved with the same set of coefficients in the matrix $\mathbf{A}$ but with different right-hand sides. In this case the correct organization of the method will save considerably on the computational time required. If Eqns (4.7), (4.8), (4.11), and (4.12), which give the computational steps for reduction to triangular form, are examined it can be seen that these calculations are independent of the right-hand side terms. After this reduction process has been done once, it need not be repeated provided the multipliers m_{ij} are stored. If all the right-hand sides are available initially it is possible to process all of these simultaneously as the reduction to triangular form is done. Otherwise, the storage of the multipliers enables subsequent right-hand sides to be solved with the minimum of effort.

The remaining equations, (4.9) and (4.13), depend only on the terms b_i and the multipliers m_{ij} which are stored when the triangular reduction is first performed. Therefore, only the calculations in Eqns (4.9) and (4.13) are carried out for subsequent right-hand sides. It is convenient that the number of zero elements which are introduced is exactly equal to the number of multipliers so that instead of storing the value zero the storage space in the computer is used to store the value of m_{ij}. For example, in the first column the multiplier which is stored in row i is m_{i1} and generally, position i,j contains m_{ij} for $j < i$.

4.2.3 Finding the inverse

A particular case, of a set of equations with several right-hand sides, occurs when it is required to find the inverse of a non-singular matrix $\mathbf{A}$. This is given by the solution $\mathbf{X}$ of the equation $\mathbf{AX} = \mathbf{I}$ where the matrix $\mathbf{A}$ has the coefficients of the left-hand side of Eqn (4.4) and the right-hand side is given by

$$
I = \begin{bmatrix}
1 & 0 & 0 & & & & 0 \\
0 & 1 & 0 & & & & \cdot \\
\cdot & 0 & 1 & & & & \cdot \\
\cdot & \cdot & 0 & & & & \cdot \\
\cdot & & \cdot & & & & \cdot \\
\cdot & & \cdot & & & & 0 \\
0 & 0 & 0 & \cdot & \cdot & 0 & 1
\end{bmatrix}
\qquad (4.15)
$$

Solving for each of these right-hand sides in turn will give the columns of the inverse matrix A^{-1}. It should be pointed out that only in rare cases is it necessary to find the matrix A^{-1} since it is not required for solving a set of equations, the latter being best effected by gaussian elimination. Perhaps it should also be pointed out that Cramer's determinant method for solving equations and the formula $A^{-1} = \text{Adj} \, A / \det(A)$ are quite unsuitable for computer use. The determinant of a matrix is found quite simply as a by-product of the gaussian elimination process. If the process is carried out as described above, then the determinant is the product of the diagonal elements of the triangular matrix which is used for the back-substitution process.

4.2.4 Partial pivoting

It can be seen that for large matrices a considerable number of arithmetic operations is involved, and at each stage in the process the calculations use the numbers calculated in the previous stage. This is the classic situation in which error magnification can take place and therefore it is important to take all possible steps to minimize the build-up of errors.

From Eqns (4.12) and (4.13) we see that one operation which occurs many times is multiplication by m_{ij}. In multiplying the number any accumulated error which is present will also be multiplied by m_{ij}; therefore, these multipliers should be made as small as possible, and certainly less than one, so that the errors are not magnified by the multiplication.

This is easily achieved if the 'pivotal' element $a_{kk}^{(k)}$ is the largest of all the elements $a_{ik}^{(k)}$ in the same column for $i \geqslant k$ since then

$$
|m_{ij}| \leqslant 1, \qquad \begin{aligned} i &= 1, 2, \ldots, n \\ j &< i \end{aligned} \qquad (4.16)
$$

To implement this suggestion requires an extra step which involves very little computation. At the stage in the process where the next x_k is about to be eliminated a search is made, down the leading column from the diagonal element downwards, and the element of largest modulus is found. The row containing this element is then interchanged with the pivotal row so that the largest element is now in the pivotal position. This gives multipliers of modulus less than one.

This is done at every stage in the process interchanging where necessary. This interchange process will also deal with the problem which would occur if a pivotal element $a_{kk}^{(k)}$ was zero. Unless the matrix is singular, or nearly singular, the search process will find at least one nonzero element in each column so that the problem of division by zero will not occur. It should be realized that this partial-pivoting process is essential to maintain accuracy if the problem is at all ill-conditioned.

It is also possible to extend this idea and apply complete pivoting to the matrix. This refinement consists of searching the whole of the remaining sub-matrix to put the element of largest modulus in the pivotal position. This may involve altering not only the order of the rows but also the order of the variables in the equations, this being more complicated to program than partial pivoting. Since the gain in accuracy over partial pivoting is not very significant the complete pivoting strategy is not often used.

4.2.5 Scaling

With a little thought it can be seen that the partial-pivoting strategy on its own is inadequate. If we consider the two equations

$$4x + 3y = 10 \qquad\qquad\qquad\qquad\qquad (4.17)$$
$$3x - 2y = 12$$

then the pivotal element is already the largest in modulus. However, the simple operation of multiplying the second equation by 2 would mean that the two equations would need to be interchanged to put the element of largest modulus in the pivotal position. Thus the choice of pivotal row is arbitrary.

It has been found that the arbitrary nature of the choice of pivotal row is a hindrance to the development of the most accurate elimination pro-cedure. The solution to this problem is to scale the matrix so that the rows are comparable in some defined way. This is usually done by normalizing in one of two ways. The rows can either be normalized by dividing the whole row by the element in the row which has largest modulus so that the largest element of the new row is one, or alternatively, each row can be divided by

$$d_i = \sqrt{\left/ \left(\sum_{j=1}^{n} a_{ij}^2 \right) \right.}$$

Although it is established that scaling can make a significant difference to the accuracy of the solutions there is no standard method of scaling which is universally accepted. The interested reader is referred to the work of Bauer (1963) and Wilkinson (1961) in which this topic is considered.

4.2.6 Ill-conditioning

It is, of course, possible that the scheme will break down even if partial pivoting is employed but the gaussian-elimination scheme will give some indication as to the type of ill-conditioning which has caused the problem. This can be illustrated by simple equations in two variables.

$$
\begin{array}{ll}
2x + 3y = 10 & 2x + 3y = 10 \\
\dfrac{4x + 6y = 20}{0 \ = 0} & \dfrac{4x + 6y = 22}{0 \ = 2}
\end{array}
\qquad (4.18)
$$

In the first example the elimination process would show a very small element not only in the pivotal position but also on the right-hand side of the equation. This would indicate that the n simultaneous equations were not linearly independent.

In the second case the right-hand side is not close to zero. This indicates that the equations are inconsistent and some mistake must have been made in the problem formulation. It is clear that at each stage the pivotal element must be checked and some appropriate action taken if the size of the pivotal element falls below some chosen small quantity.

4.2.7 Improvement of the solution

As an elementary precaution, when the solution of the set of equations has been found the values should be inserted into the original equations to see how closely they are satisfied. If there are large errors then clearly the solutions are not satisfactory. Unfortunately, with ill-conditioned equations quite small errors may be found after the substitutions, even though the computed solutions differ considerably from the true solution.

Let the computed solutions be

$$\mathbf{X}^{(0)} \equiv [x_1^{(0)}, x_2^{(0)}, \ldots, x_n^{(0)}]^T \tag{4.19}$$

We form the residuals in double precision

$$\mathbf{R}^{(0)} \equiv [r_1^{(0)}, r_2^{(0)}, \ldots, r_n^{(0)}]^T \tag{4.20}$$

by substituting in the equations (4.4). In matrix form

$$\mathbf{AX}^{(0)} - \mathbf{B} = \mathbf{R}^{(0)} \tag{4.21}$$

The residuals $\mathbf{R}^{(0)}$ will indicate when the solutions are unacceptable but they also lead to a method of improving the solution which is often quite effective. Since, the true solution satisfies

$$\mathbf{AX} - \mathbf{B} = 0 \tag{4.22}$$

then subtracting Eqn (4.21) from Eqn (4.22) gives

$$\mathbf{A}(\mathbf{X} - \mathbf{X}^{(0)}) = -\mathbf{R}^{(0)} \tag{4.23}$$

The quantity $\mathbf{X} - \mathbf{X}^{(0)}$ is the quantity which should be added to the first computed solution $\mathbf{X}^{(0)}$ in order to obtain the correct solution. If this could be calculated exactly the problem is completely solved. In fact, $\mathbf{X} - \mathbf{X}^{(0)}$ is found as before by solving the set of linear simultaneous equations. It should be noted however that the work is considerably reduced because the reduction to triangular form and the storage of the multipliers has been performed.

It has been found that solving Eqn (4.23) and using this solution to correct the value $\mathbf{X}^{(0)}$ often increases the accuracy considerably and at least one improvement of this type is recommended in cases where there is concern for accuracy. Formally, we let $\mathbf{E}^{(0)}$ be the solution of $\mathbf{AE} = -\mathbf{R}^{(0)}$ and the new approximation to the solution is given by $\mathbf{X}^{(1)} = \mathbf{X}^{(0)} + \mathbf{E}^{(0)}$.

Further improvements can be made by forming the residuals

$$\mathbf{AX}^{(p)} - \mathbf{B} = \mathbf{R}^{(p)} \tag{4.24}$$

in double precision and then solving a succession of equations

$$\mathbf{AE} = -\mathbf{R}^{(p)}, \quad p = 1, 2, \ldots \tag{4.25}$$

The solutions $\mathbf{E}^{(p)}$ of these equations give the new approximations to the solutions

$$\mathbf{X}^{(p+1)} = \mathbf{X}^{(p)} + \mathbf{E}^{(p)} \tag{4.26}$$

This procedure is performed normally no more than once or twice since further work gives little extra improvement.

4.2.8   Other direct methods

There are several variants of gaussian elimination which are described in various texts. For example, a scheme which was popular for desk-machine work is the Jordan elimination scheme. In this variation the final form of the matrix after elimination is a diagonal form. Each equation has only one variable, therefore, the back-substitution process is avoided and the values of the variables can be calculated directly.

The method proceeds in a very similar manner to the gaussian elimination method but at each stage the variable x_k is eliminated, not only from the succeeding equations but also from all the preceding ones. Equations (4.12) and (4.13) are used for all $i \neq k$. Jordan elimination needs approximately $n^3/2$ operations compared with $n^3/3$ for gaussian elimination and is, therefore, not recommended for general use.

There is another group of methods which can be described under the general heading of triangular decomposition; these include the variants of Crout, Doolittle, and Choleski. The computational scheme is presented in a different manner from that of gaussian elimination although the schemes are very similar. It is based on a series of multipliers which reduce the matrix to triangular form followed by the process of back-substitution.

The scheme does have an advantage over gaussian elimination on those computers which have the facility for accumulating products of the form $\sum_{j=1}^{n} a_j b_j$ in double length. The attraction of this facility is that it does not incur the time penalty associated with normal double-length working. A multiplication on a computer always involves the production of a double-length number before truncating to continue the arithmetic. On the machines in which we are interested the truncation back to single length only occurs after the addition of the products has taken place. Since gaussian elimination does not present the arithmetic in the form of sums of products the triangular-decomposition method is to be preferred where the double-length accumulation is available.

It is assumed that the triangular decomposition is theoretically possible, i.e., the matrix can be expressed as the product of two matrices

$$\mathbf{A} = \mathbf{L} . \mathbf{U} \tag{4.27}$$

where $\mathbf{U}$ is an upper-triangular matrix and $\mathbf{L}$ is a lower-triangular matrix. Once these matrices have been found the set of equations is solved in two stages. Each of these involves the solution of a set of equations with a triangular matrix, this being simple to perform by a process of forward- or back-substitution. The equation becomes

$$\mathbf{L} . \mathbf{U} . \mathbf{X} = \mathbf{B} \tag{4.28}$$

and we find a vector $\mathbf{Y}$ such that $\mathbf{L} . \mathbf{Y} = \mathbf{B}$ and then solve the equations $\mathbf{U} . \mathbf{X} = \mathbf{Y}$. It is possible to introduce pivoting, as in the case of gaussian elimination, and this would normally be done. However, it has been shown by Wilkinson (1961) that for a symmetric positive definite matrix the errors are not significantly increased by omitting the partial pivoting.

The method will be illustrated by means of a set of equations of dimension 3. We must find coefficients such that

$$\begin{pmatrix} l_{11} & 0 & 0 \\ l_{21} & l_{22} & 0 \\ l_{31} & l_{32} & l_{32} \end{pmatrix}\begin{pmatrix} 1 & u_{12} & u_{13} \\ 0 & 1 & u_{23} \\ 0 & 0 & 1 \end{pmatrix} = \begin{pmatrix} a_{11} & a_{12} & a_{13} \\ a_{21} & a_{22} & a_{23} \\ a_{31} & a_{32} & a_{33} \end{pmatrix} \tag{4.28}$$

There is an element of freedom in the choice of the diagonal values of L and U and in this case the matrix U has been given diagonal elements equal to 1 for simplicity.

By equating the coefficients of the right- and left-hand sides of Eqn (4.28) the coefficients of L and U can be found. The equations for the first column of A give the values l_{11}, l_{21}, and l_{31} directly.

$$l_{11} = a_{11}, \qquad l_{21} = a_{21}, \qquad l_{31} = a_{31} \tag{4.30}$$

The equations for the second column can be solved for u_{12}, l_{22}, and l_{32}

$$l_{11}u_{12} = a_{12}, \qquad l_{21}u_{12} + l_{22} = a_{22}, \qquad l_{31}u_{12} + l_{32} = a_{32} \tag{4.31}$$

and the equation for the third column gives the values of u_{13}, u_{23}, and l_{33}

$$\begin{aligned} l_{11}u_{13} &= a_{13} \\ l_{21}u_{13} + l_{22}u_{23} &= a_{23} \\ l_{31}u_{13} + l_{32}u_{23} + l_{33} &= a_{33} \end{aligned} \tag{4.32}$$

The general form of this algorithm is described in Isaacson and Keller (1966). The number of operations is the same as for gaussian elimination and, in view of the possibility of accumulating products double length, this is the method recommended for general use where a direct method is required.

In the case of a symmetric matrix it is possible to reduce the amount of computation and storage by taking advantage of the symmetry. If the diagonal elements of L and U are made equal then $U = L^T$ and only the elements of L need be calculated or stored. This is the method known as Choleski factorization.

The method of triangular decomposition is mentioned again in the chapter on finding eigenvalues where it forms the basis of a method for obtaining all the eigenvalues of a matrix. A numerical example is given in Example 4.4.

4.2.9 Some special matrix forms
A matrix form which frequently arises is the tridiagonal matrix which has the form

$$\begin{bmatrix} a_1 & c_1 & & & & & \\ b_2 & a_2 & c_2 & & & 0 & \\ & & & \ddots & & & \\ & & & & b_{n-1} & a_{n-1} & c_{n-1} \\ & 0 & & & & b_n & a_n \end{bmatrix} \tag{4.33}$$

All the elements are zero except those on the main diagonal and the one above and below. Such a matrix could occur in the solution of boundary-value problems for ordinary differential equations in which case $|a_i| \geqslant |b_i| + |c_i|$.

It will be seen later that this condition ensures that the iterative methods of the next section can be used. Also, since these iterative methods are particularly suitable for matrices with a large number of zero elements, one may well be tempted to use iterative methods for this type of equation. In fact, because the elements are grouped so closely round the diagonal, it is possible to develop a variant of gaussian elimination which only calculates with the nonzero elements and takes full advantage of the sparsity. Since the work involved for an iterative method depends on the number of iterations the direct method is to be preferred in this instance.

If the matrix equation is $\mathbf{AX} = \mathbf{V}$ then the equations for reduction to triangular form in this algorithm, which is known as the Thomas algorithm, are

$$\alpha_1 = a_1 \qquad \gamma_1 = c_1/\alpha_1 \qquad u_1 = v_1/\alpha_1 \tag{4.34}$$

$$\begin{aligned} \alpha_i &= a_i - b_i\gamma_{i-1}, & i &= 2, 3, \ldots, n \\ u_i &= (v_i - b_iu_{i-1})/\alpha_i, & i &= 2, 3, \ldots, n \\ \gamma_i &= c_i/\alpha_i & i &= 2, 3, \ldots, n-1 \end{aligned} \tag{4.35}$$

and the back-substitution solution is given by

$$\begin{aligned} x_n &= u_n, \\ x_i &= u_i - \gamma_ix_{i+1}, & i &= n-1, n-2, \ldots, 1 \end{aligned} \tag{4.36}$$

Numerical results for this method are given in Example 4.5.

The advantage of formulating the computation in this way is, that for certain commonly occurring sets of a_i, b_i, and c_i, it can be shown that the multipliers γ_i have modulus less than unity which gives the same effect as was achieved by pivoting in ordinary gaussian elimination. This is discussed in Isaacson and Keller (1966). The striking fact about the above algorithm is that the number of operations required is approximately equal to $5n$ for large n as compared with $n^3/3$ for gaussian elimination on a full matrix. It should also be noted that a similar scheme can be devised for matrices with elements grouped round the diagonal where the bandwidth is greater than 3.

Another commonly occurring form of matrix is the block tridiagonal matrix which occurs in the finite-difference solution of partial-differential equations. The form is similar to that of Eqn (4.33) but in this case the elements are matrices. Equations (4.34), (4.35), and (4.36) for the reduction to triangular form can be used, provided the matrix quantity α_i^{-1} replaces division by α_i. The process does involve the calculation of an inverse matrix but this can be done economically by gaussian elimination. This method is considerably more efficient than simple iteration or full gaussian elimination. Further details are available in Smith (1965).

4.3 Iterative methods

4.3.1 Jacobi's method

We have seen in Chapter 2 that it is possible to use an iterative scheme which improves the solution until convergence is obtained. This idea can be very easily applied to linear simultaneous equations as is shown by the following example.

Let us write the following equations

$$
\begin{aligned}
10x_1 + x_2 + x_3 &= 24 \\
-x_1 + 20x_2 + x_3 &= 21 \\
x_1 - 2x_2 + 100x_3 &= 300
\end{aligned}
\tag{4.37}
$$

in the form

$$
\begin{aligned}
x_1 &= \tfrac{1}{10}(24 - x_2 - x_3) \\
x_2 &= \tfrac{1}{20}(21 + x_1 - x_3) \\
x_3 &= \tfrac{1}{100}(300 - x_1 + 2x_2)
\end{aligned}
\tag{4.38}
$$

If the first approximations are $(0,0,0)$ then the next iterate is $(2\cdot4, 1\cdot05, 3\cdot00)$ and successive iterates are $(1\cdot995, 1\cdot02, 2\cdot997)$, $(1\cdot9983, 0\cdot9999, 2\cdot999\,55)$. Substitution of the values $(2, 1, 3)$ in the equations shows that the process is rapidly converging to the true solution.

It is certainly clear that iteration could be a very useful method of solution: we now show that it can also be a very unsuitable method in certain circumstances. Equations (4.37) can be taken in a different order to give

$$
\begin{aligned}
x_1 &= 300 + 2x_2 - 100x_3 \\
x_2 &= 24 - 10x_1 - x_3 \\
x_3 &= 21 + x_1 - 20x_2
\end{aligned}
\tag{4.39}
$$

Using the values $(0,0,0)$ for the first approximation gives successive approximations $(300, 24, 21)$ and $(-1752, -2997, -3879)$ and there is no possibility of this sequence converging to the values $(2, 1, 3)$. It is crucial that some knowledge is obtained of conditions which will ensure that the iteration converges. This matter is discussed later in the chapter.

The above method which is known as Jacobi's method is rarely used because there are various improvements which can be used to speed up the rate of convergence. However, in certain circumstances the other methods may not converge and possibly the Jacobi method may be suitable.

4.3.2 Gauss–Seidel method

In studying the simple example above the reader may have wondered why a block of values are found and substituted together into the equations. It seems reasonable that as new values are found they are substituted immediately into subsequent equations. This idea is the basis of the Gauss–Seidel method which can effect a considerable improvement in the rate of convergence for certain types of equations.

In order to see the way in which the different methods work it is convenient to split the coefficients into three groups; namely, the set of diagonal elements, the elements above the diagonal, and the elements below the diagonal.

The most convenient way of expressing these methods is to use matrix notation. The matrix **A** is split into three parts corresponding to the three sets of coefficients discussed above. The equation

$$\mathbf{AX} = \mathbf{B}$$

becomes

$$(\mathbf{L} + \mathbf{D} + \mathbf{U})\mathbf{X} = \mathbf{B} \tag{4.40}$$

It will be convenient to scale the equations by dividing through by the diagonal elements so that **D** becomes equal to the unit matrix **I**.

The Jacobi method results from transferring all terms to the right-hand side except the diagonal terms, and iterating as follows

$$\mathbf{X}^{(r+1)} = (-\mathbf{L} - \mathbf{U})\mathbf{X}^{(r)} + \mathbf{B}, \qquad r = 0, 1, \ldots \tag{4.41}$$

However, the Gauss–Seidel method introduces $x_1^{(r+1)}, x_2^{(r+1)}$ etc., on the right-hand side as soon as they are available. Thus, the iteration equations become

$$\mathbf{X}^{(r+1)} = -\mathbf{L} \cdot \mathbf{X}^{(r+1)} - \mathbf{U}\mathbf{X}^{(r)} + \mathbf{B}$$

or

$$(\mathbf{I} + \mathbf{L})\mathbf{X}^{(r+1)} = -\mathbf{U}\mathbf{X}^{(r)} + \mathbf{B} \tag{4.42}$$

Writing these equations in full

$$x_1^{(r+1)} = -(a_{12}x_2^{(r)} + a_{13}x_3^{(r)} + \cdots + a_{1n}x_n^{(r)}) + b_1$$
$$x_2^{(r+1)} = -(a_{21}x_1^{(r+1)} + a_{23}x_3^{(r)} + \cdots + a_{2n}x_n^{(r)}) + b_2$$
$$x_3^{(r+1)} = -(a_{31}x_1^{(r+1)} + a_{32}x_2^{(r+1)} + \cdots + a_{3n}x_n^{(r)}) + b_3 \tag{4.43}$$
$$\cdots\cdots\cdots\cdots\cdots\cdots\cdots\cdots\cdots\cdots\cdots\cdots$$
$$x_n^{(r+1)} = -(a_{n1}x_1^{(r+1)} + a_{n2}x_2^{(r+1)} + \cdots + a_{n,n-1}x_{n-2}^{(r+1)}) + b_n$$

When both the Jacobi and Gauss–Seidel methods converge it can be shown that the Gauss–Seidel method gives faster convergence than the Jacobi method. Example 4.3 shows numerical results for Jacobi, Gauss–Seidel, and the over-relaxation method discussed below. The difference in the rates of convergence is quite striking.

4.3.3 Over relaxation

In chapter 1 it was shown that if an iterative process is converging slowly, it is sometimes possible to take larger steps than the calculated value, thus accelerating the convergence. This technique is almost always adopted for the linear simultaneous equations which arise in the solution of elliptic differential equations. These equations satisfy conditions which guarantee convergence but the rate of convergence, particularly for a large system, is very slow. The values calculated from the Gauss–Seidel process are therefore looked upon as intermediate values and the following equation is used to find the modified values.

$$\mathbf{X}^{(r+1)} = \mathbf{X}^{(r)} + \omega(\breve{\mathbf{X}}^{(r+1)} - \mathbf{X}^{(r)}) \tag{4.44}$$

$\breve{\mathbf{X}}^{(r+1)}$ is the value calculated by the Gauss–Seidel process. It can be seen from the equation that the new value is obtained by multiplying the increment from the last approximation by a factor ω. When $\omega > 1$ we have an over-relaxation method whereas if $\omega < 1$ the system of equations is said to be under-relaxed. For elliptic equations the value ω normally satisfies $1 < \omega < 2$. In matrix terms the equation becomes

$$X^{(r+1)} = X^{(r)} + \omega(-LX^{(r+1)} - UX^{(r)} + B - X^{(r)})$$
$$= [-\omega U + (1 - \omega)I]X^{(r)} - \omega LX^{(r+1)} + \omega B \qquad (4.45)$$

Clearly, the choice of ω is the main difficulty in using the over-relaxation method. However, the matrices which arise in the solution of partial-differential equations often have particularly simple forms and for some of these it is possible to calculate the most effective value of ω. In particular it emerges that an over-estimate of ω is preferable to an under-estimate. In the case of a more general matrix it would be feasible to use over-relaxation if it were possible to conduct a series of experiments, with different relaxation factors to determine the most suitable values.

4.3.4 Convergence of iterative methods

In order to investigate the various methods we will write the above equations in the general form

$$X^{(r+1)} = PX^{(r)} + C \qquad (4.46)$$

The final solution X is given by the equation

$$X = PX + C \qquad (4.47)$$

and, therefore, the error $E^{(r)} = X - X^{(r)}$ is given by

$$X - X^{(r+1)} = P(X - X^{(r)})$$
$$E^{(r+1)} \quad\;\; = PE^{(r)} \qquad (4.48)$$
$$= P^{r+1}E^{(0)}$$

It is clear that convergence will only be achieved if the effect of the matrix P is to successively reduce the error $E^{(r)}$. One condition which is clearly necessary is that all the eigenvalues of P shall have modulus less than one. Otherwise, if we have $|\lambda_i| > 1$ with a corresponding eigenvector v_i then letting $E^{(0)} = v_i$

$$E^{(r+1)} = P^{r+1}v_i$$
$$= \lambda^{r+1}v_i \qquad (4.49)$$

which grows without bound as r increases. The condition $|\lambda_i| < 1$ is also a suffi-cient condition and, therefore, a knowledge of the eigenvalues of P will determine whether the iteration will converge. A definition of the terms *eigenvalue* and *eigenvector* is given in Chapter 9.

The eigenvalues themselves are difficult to evaluate but there are various conditions which can easily be checked which will give an upper bound for the modulus of the eigenvalues. If such an upper bound is less than unity then the process will converge. However, if these upper bounds are greater than unity then it is still possible for the eigenvalues to have modulus less than unity and so the condition on the upper bound is not necessary for convergence.

In the first two iterative processes considered above the matrix $\mathbf{P}$ has the forms

Jacobi	$-\mathbf{L}-\mathbf{U}$	(4.50)
Gauss–Seidel	$(\mathbf{I}+\mathbf{L})^{-1}(-\mathbf{U})$	(4.51)

It can be shown that the condition of diagonal dominance of the original matrix $\mathbf{A}$ is sufficient to ensure the convergence of the processes represented by the above matrices. (See, for example, Isaacson and Keller (1966).) A matrix is said to be strictly diagonally dominant if

$$d_r < 1 \quad \text{for} \quad r = 1, 2, \ldots, n \tag{4.52}$$

where

$$d_r = \frac{\displaystyle\sum_{j=1}^{n}{}' |a_{r,j}|}{|a_{rr}|} \tag{4.53}$$

with the prime notation signifying that the value a_{rr} is omitted from the summation. If $d_r \leqslant 1$ for $r = 1, 2, \ldots, n$ and $d_r < 1$ for at least one value of r, then the matrix is said to be weakly diagonally dominant. This condition is sufficient for convergence of the iterative process.

Another condition which will ensure convergence is when the matrix $\mathbf{A}$ is positive definite. Such matrices quite often occur in physical problems and in the solution of least-squares problems. A matrix is said to be positive definite if for every non-null vector $\mathbf{X}$ the quantity $\mathbf{X}^T \mathbf{A} \mathbf{X} > 0$. Since this property is more difficult to investigate, the property of diagonal dominance is more often used to check if convergence can be guaranteed. In the case of over-relaxation it is possible to give definite conclusions about convergence for certain matrices with special forms. The interested reader is referred to the book by Varga (1962) where these problems are discussed in a rigorous and detailed manner.

4.4 Sparse matrices

The term sparse is used to describe a matrix which has a large number of zero elements. For this type of matrix, the amount of computation time required for the various direct or iterative schemes does not follow the normal pattern. An iterative method normally requires mn^2 operations for its solution where m is the number of iterations and n the number of equations. However, if the method is programmed to include in the calculations only the nonzero elements, then the amount of computation is proportional to the number of nonzero elements. This can reduce the computational time to the level where iterative methods become more economic than the direct methods. Therefore, for a randomly sparse matrix with no special structure an iterative method would probably be used. There are two situations where this would not be the case; namely, if the matrix is not diagonally dominant then a direct method should be used to avoid convergence difficulties, or if the matrix has a special structure then special algorithms can often be designed to exploit this.

In the case of random sparsity where the number of nonzero elements in the matrix initially is small, say 5% or 10%, there are some interesting variations on the standard gaussian elimination method. The problem which arises in the standard method is that each subtraction of rows introduces further nonzero elements, and it is possible for the initially very sparse matrix to fill up rapidly. Various strategies have been considered to reduce the occurrence of these extra nonzero elements in programs which are designed to calculate only with the nonzero elements.

A simple strategy which has given good experimental results is to use a new type of pivoting strategy. At each stage in the process, the pivotal row is chosen as the one with the smallest number of nonzero elements which also has a satisfactory element in the pivotal position. It is, of course, important that the element in the pivotal position is not too small, since with the strategy outlined here the pivotal element will not generally be the largest as it was with the previous gaussian method.

The other exception is when the matrix has a simple form, which allows a more efficient form of gaussian elimination to be adopted, so that the amount of computation time is no longer proportional to $n^3/3$. This is exemplified by the Thomas algorithm, which is a reformulation of the gaussian elimination scheme applied to a tridiagonal matrix, i.e.,

$$
\begin{bmatrix}
a_1 & c_1 & & & & \\
b_2 & a_2 & c_2 & & 0 & \\
& & & & & \\
& 0 & & b_{n-1} & a_{n-1} & c_{n-1} \\
& & & & b_n & a_n
\end{bmatrix}
\tag{4.54}
$$

This is described in Section 4.2.

4.5 Comparison of methods

The computational time required by each method is an important consideration and details of the operational counts for the various methods are given in Table 4.1. On many computers the times for multiplication and division are much longer than those for addition and subtraction and, therefore, the operational count is for multiplication and divisions only.

Table 4.1
Operation counts for standard methods

Gaussian elimination $\Big\}$ Triangular decomposition	$\dfrac{n^3}{3}+n^2-\dfrac{n}{3}$
Invert matrix and multiply	n^3+n^2
Jordan elimination	$\dfrac{n^3}{2}+n^2-\dfrac{n}{2}$
Further right-hand side for all above methods	n^2
Tridiagonal solution by Thomas algorithm	$5n-4$
Further right-hand side	$3n-2$
Iterative methods	rn^2
Further right-hand side	$r.n^2$

n is the number of equations.
r is the number of iterations required for some specified convergence criteria.

In the class of direct methods it is clear that those methods which exploit any special structure in the matrix give considerable economies over the standard methods. For a general matrix which is to be solved by direct methods either gaussian elimination or triangular decomposition would be used. If a computer is available with the facility to accumulate products double-length then the triangular-decomposition algorithm will give more accurate results. The triangular decomposition method is also useful when the matrix is symmetric since it is then possible to reduce the number of operations and cut down the storage requirements by 50%.

The iterative methods are used for matrices with random sparse elements where the rate of convergence is good. If the method converges slowly for any problem it may be preferable to use direct methods, even though the matrix is sparse. The method of pivoting according to the row with the smallest number of nonzero elements has given some encouraging experimental results, however, there is the danger that errors may accumulate owing to the modification of the normal pivoting policy.

In the absence of any other guideline the iterative method which would normally be used is the Gauss–Seidel method since this usually has a faster rate of convergence than Jacobi's method. If the matrix has a special form, such as those arising in some partial-differential equations, then it may be possible to calculate a suitable over-relaxation factor ω which would then be used in order to accelerate the convergence. For a general matrix the choice of an over-relaxation factor is difficult, but it may be worthwhile if the rate of convergence is slow and if several similar calculations are to be performed. A series of calculations with trial values of ω would be performed and a suitable value of ω would then be chosen.

If the Gauss–Seidel method fails to converge then the alternative Jacobi method is available.

Bibliographic notes

The book by McCracken and Dorn (1964) provides a simple, but fairly detailed treatment, of many of the methods. The full equations for the eliminations together with flow diagrams of the method and worked examples are given. Similar material is presented in Conte (1965) from a rather more mathematical viewpoint; some of the important mathematical properties of matrices are discussed as well as the computational methods. Flow diagrams and FORTRAN programs are provided.

Brief descriptions of the method together with several worked examples and ALGOL programs are given in Bull (1966) but little theoretical background is presented.

A detailed text which does not require specialist mathematical training is provided by the book of Fox (1964).

The reader interested in a more extensive error analysis of the direct methods will find the text by Fox and Mayers (1968) helpful. Very little attention is given in this book to iterative methods.

A good mathematical treatment is provided by the books of Isaacson and Keller (1966) and Ralston (1965) which give both a description of the computational procedure and error analysis of the methods. Ralston (1965) also contains a section on methods which have not been discussed here.

There are many specialized books which the mathematician who is interested in this area of study will find worthwhile. The book by Varga (1962) provides a careful and detailed analysis of the convergence problems of iterative methods, this being the source text for many of the quoted results. The various texts by Wilkinson (1961), (1964), etc. are an important source of many of the fundamental results, particularly in the field of error analysis. The text by Forsythe and Moler (1967) gives a detailed treatment of direct-solution methods which is accessible to the reader with mathematical training below the specialist level.

Worked examples

1 The table of figures on page 74 shows a gaussian elimination scheme for four equations in four unknowns. The multipliers form a lower triangle which can be stored in the positions in which zero elements are produced by the gaussian elimination process. The upper-triangular matrix used for back-substitution is formed by taking the top row of each set of equations. The last row of figures gives the values for x_4, x_3, x_2, and x_1 which are found in the back-substitution process.

    Each set of equations is found as described in Section 4.2.1 by subtracting the appropriate multiple of the top equation of a set. This generates the new set of equations of dimension one less than the previous set.

2 An example of an ill-conditioned matrix is the Hilbert segment matrix.

$$\begin{bmatrix} 1 & \tfrac{1}{2} & \tfrac{1}{3} & \tfrac{1}{4} \\ \tfrac{1}{2} & \tfrac{1}{3} & \tfrac{1}{4} & \tfrac{1}{5} \\ \tfrac{1}{3} & \tfrac{1}{4} & \tfrac{1}{5} & \tfrac{1}{6} \\ \tfrac{1}{4} & \tfrac{1}{5} & \tfrac{1}{6} & \tfrac{1}{7} \end{bmatrix}$$

The results of inverting this matrix with six significant figures and with three significant figures are presented in Example 7.6.

3 The table of results shown on pp. 75 and 76 gives an interesting comparison between the various iterative methods for the solution of linear simultaneous equations. The matrix equation to be solved is $AX = B$ where

$$A = \begin{bmatrix} 1\cdot0 & -0\cdot7 & 0\cdot0 \\ -0\cdot7 & 1\cdot0 & -0\cdot7 \\ 0\cdot0 & -0\cdot7 & 1\cdot0 \end{bmatrix} \qquad B = \begin{bmatrix} -4 \\ 34 \\ -44 \end{bmatrix}$$

    The value of the largest modulus of all the eigenvalues determines the rate of convergence. Ideally, this value should be substantially less than unity although the theoretical condition simply requires a value less than unity. The value for this quantity (the spectral radius) is 0·9898 for the Jacobi method, 0·98 for the Gauss–Seidel method, and 0·755 for over-relaxation with the optimum relaxation factor of (in this case) 1·755. It can be seen that the Jacobi method is still quite far from the true solution after 500 iterations whereas the over-relaxation method has converged satisfactorily in 70 iterations.

    This example was provided by Dr R. W. Thatcher of the Mathematics Department at UMIST.

Worked example 1

Multipliers					Right hand side
	1·400 000 0000	0·350 000 0000	0·980 000 0000	0·029 224 0000	2·639 600 0001
0·250 000 0000	0·350 000 0000	0·980 000 0000	0·029 224 0000	0·006 633 0000	0·669 855 0000
0·700 000 0000	0·980 000 0000	0·029 224 0000	0·006 633 0000	0·002 887 0000	0·189 451 0000
0·020 874 2857	0·029 224 0000	0·006 633 0000	0·002 887 0000	0·000 935 0000	0·056 897 0000

Multipliers				
	0·892 500 0000	−0·215 776 0000	−0·000 673 0000	−0·009 955 0000
−0·241 765 8263	−0·215 776 0000	−0·679 367 0000	−0·017 569 8000	−1·658 269 0001
−0·000 754 0616	−0·000 673 0000	−0·017 569 8000	0·000 324 9699	0·001 797 2354

Multiplier			
		−0·731 534 2630	−0·017 732 5084	−1·655 862 2213
0·024 240 1611	−0·017 732 5084	0·000 324 4624	0·001 804 7421

| | | | 0·000 754 3013 | 0·041 943 1092 |

x_1	x_2	x_3	x_4
55·605 249 3164	0·915 666 8450	0·274 460 7977	0·015 126 7191

Jacobi's method

Iteration number	x_1	x_2	x_3	Iteration number	x_1	x_2	x_3
0	0·0000	0·0000	0·0000	300	10·4830	19·0341	−29·5170
1	−4·0000	34·0000	−44·0000	301	9·3239	20·6761	−30·6761
2	19·8000	0·4000	−20·2000	302	10·4733	19·0534	−29·5267
3	−3·7200	33·7200	−43·7200	303	9·3374	20·6626	−30·6626
4	19·6040	0·7920	−20·3960	304	10·4638	19·0723	−29·5362
5	−3·4456	33·4456	−43·4456	305	9·3506	20·6494	−30·6494
6	19·4119	1·1762	−20·5881	306	10·4546	19·0909	−29·5454
7	−3·1767	33·1767	−43·1767	307	9·3636	20·6364	−30·6364
8	19·2237	1·5526	−20·7763	308	10·4455	19·1091	−29·5545
9	−2·9132	32·9132	−42·9132	309	9·3763	20·6237	−30·6237
100	13·6417	12·7166	−26·3583	400	10·1759	19·6482	−29·8241
101	4·9016	25·0984	−35·0984	401	9·7538	20·2462	−30·2462
102	13·5689	12·8623	−26·4311	402	10·1724	19·6553	−29·8276
103	5·0036	24·9964	−34·9964	403	9·7587	20·2413	−30·2413
104	13·4975	13·0050	−26·5025	404	10·1689	19·6622	−29·8311
105	5·1035	24·8965	−34·8965	405	9·7635	20·2365	−30·2365
106	13·4275	13·1449	−26·5725	406	10·1655	19·6689	−29·8345
107	5·2014	24·7986	−34·7986	407	9·7682	20·2318	−30·2318
108	13·3590	13·2820	−26·6410	408	10·1622	19·6755	−29·8378
109	5·2974	24·7026	−34·7026	409	9·7729	20·2271	−30·2271
200	11·3262	17·3476	−28·6738	490	10·0709	19·8583	−29·9291
201	8·1433	21·8567	−31·8567	491	9·9008	20·0992	−30·0992
202	11·2997	17·4007	−28·7003	492	10·0694	19·8611	−29·9306
203	8·1805	21·8195	−31·8195	493	9·9028	20·0972	−30·0972
204	11·2737	17·4526	−28·7263	494	10·0681	19·8639	−29·9319
205	8·2169	21·7831	−31·7831	495	9·9047	20·0953	−30·0953
206	11·2482	17·5036	−28·7518	496	10·0667	19·8666	−29·9333
207	8·2525	21·7475	−31·7475	497	9·9066	20·0934	−30·0934
208	11·2232	17·5535	−28·7768	498	10·0654	19·8693	−29·9346
209	8·2875	21·7125	−31·7125	499	9·9085	20·0915	−30·0915

Gauss–Seidel method

Iteration number	x_1	x_2	x_3	Iteration number	x_1	x_2	x_3
0	0·0000	0·0000	0·0000	200	10·1436	20·2010	−29·8593
1	−4·0000	31·2000	−22·1600	201	10·1407	20·1970	−29·8621
2	17·8400	30·9760	−22·3168	202	10·1379	20·1930	−29·8649
3	17·6832	30·7565	−22·4705	203	10·1351	20·1892	−29·8676
4	17·5295	30·5414	−22·6211	204	10·1324	20·1854	−29·8702
5	17·3789	30·3305	−22·7686	205	10·1298	20·1817	−29·8728
6	17·2314	30·1239	−22·9133	206	10·1272	20·1781	−29·8754
7	17·0867	29·9214	−23·0550	207	10·1246	20·1745	−29·8779
8	16·9450	29·7230	−23·1939	208	10·1221	20·1710	−29·8803
9	16·8061	29·5285	−23·3300	209	10·1197	20·1676	−29·8827
100	11·0826	21·5157	−28·9390	290	10·0233	20·0326	−29·9772
101	11·0610	21·4853	−28·9603	291	10·0223	20·0320	−29·9776
102	11·0397	21·4556	−28·9811	292	10·0224	20·0313	−29·9781
103	11·0189	21·4265	−29·0014	293	10·0219	20·0307	−29·9785
104	10·9986	21·3980	−29·0214	294	10·0215	20·0301	−29·9789
105	10·9786	21·3700	−29·0410	295	10·0211	20·0295	−29·9794
106	10·9590	21·3426	−29·0602	296	10·0206	20·0289	−29·9798
107	10·9398	21·3158	−29·0790	297	10·0202	20·0283	−29·9802
108	10·9210	21·2895	−29·0974	298	10·0198	20·0278	−29·9806
109	10·9026	21·2637	−29·1154	299	10·0194	20·0272	−29·9810

Successive over-relaxation

$\omega = 1 \cdot 75$

Iteration number	x_1	x_2	x_3	Iteration number	x_1	x_2	x_3
0	0·0000	0·0000	0·0000	35	10·0969	20·1232	−29·9238
1	−7·0000	50·9250	−14·6169	36	10·0782	20·0967	−29·9387
2	60·6331	77·6762	29·1159	37	10·0598	20·0759	−29·9530
3	42·6784	89·1910	10·4220	38	10·0481	20·0595	−29·9623
4	70·2502	91·4302	27·1854	39	10·0369	20·0467	−29·9711
5	52·3143	88·3146	10·7963	40	10·0295	20·0366	−29·9769
6	61·9496	82·3778	15·8156	41	10·0227	20·0287	−29·9822
7	47·4506	75·2177	3·2800	42	10·0181	20·0225	−29·9858
8	49·5538	67·8081	3·6049	43	10·0139	20·0176	−29·9891
9	38·8996	60·7120	−5·3315	44	10·0111	20·0138	−29·9913
10	38·1974	54·2268	−6·5736	45	10·0086	20·0108	−29·9933
11	30·7797	48·4824	−12·6788	46	10·0068	20·0085	−29·9947
12	29·3062	43·5067	−14·1952	47	10·0052	20·0066	−29·9959
13	24·3161	39·2681	−18·2502	48	10·0042	20·0052	−29·9967
14	22·8664	35·7037	−19·5753	49	10·0032	20·0041	−29·9975
15	19·5872	32·7369	−22·2159	50	10·0026	20·0032	−29·9980
16	18·4122	30·2879	−23·2354	51	10·0020	20·0025	−29·9985
17	16·2935	28·2802	−24·9302	52	10·0016	20·0019	−29·9988
18	15·4231	26·6437	−25·6638	53	10·0012	20·0015	−29·9991
19	14·0712	25·3163	−26·7397	54	10·0010	20·0012	−29·9993
20	13·4590	24·2440	−27·2464	55	10·0007	20·0009	−29·9994
21	12·6046	23·3808	−27·9237	56	10·0006	20·0007	−29·9995
22	12·1881	22·6883	−28·2641	57	10·0005	20·0006	−29·9996
23	11·6521	22·1340	−28·6877	58	10·0004	20·0004	−29·9997
24	11·3751	21·6916	−28·9120	59	10·0003	20·0003	−29·9998
25	11·0408	21·3391	−29·1756	60	10·0002	20·0003	−29·9998
26	10·8598	21·0588	−29·3213	61	10·0002	20·0002	−29·9999
27	10·6522	20·8363	−29·4846	62	10·0001	20·0002	−29·9999
28	10·5353	20·6599	−29·5782	63	10·0001	20·0001	−29·9999
29	10·4069	20·5202	−29·6791	64	10·0001	20·0001	−29·9999
30	10·3321	20·4098	−29·7387	65	10·0001	20·0001	−30·0000
31	10·2529	20·3226	−29·8008	66	10·0000	20·0001	−30·0000
32	10·2055	20·2538	−29·8385	67	10·0000	20·0000	−30·0000
33	10·1568	20·1996	−29·8767	68	10·0000	20·0000	−30·0000
34	10·1269	20·1568	−29·9004	69	10·0000	20·0000	−30·0000

4 Factorize the following matrix into the product LU of a lower- and upper-triangular matrix.

$$\mathbf{A} = \begin{bmatrix} 4 & 2 & -1 & 0 \\ 1 & -2 & 3 & 1 \\ 2 & -3 & 5 & 1 \\ -1 & 2 & -1 & 6 \end{bmatrix}$$

Show how this enables the equation $\mathbf{AX} = \mathbf{B}$ to be solved where $\mathbf{B}^T = [0, -5, -5, 3]$
Since $\mathbf{LU} = \mathbf{A}$ we have

$$\begin{bmatrix} l_{11} & 0 & 0 & 0 \\ l_{21} & l_{22} & 0 & 0 \\ l_{31} & l_{32} & l_{33} & 0 \\ l_{41} & l_{42} & l_{43} & l_{44} \end{bmatrix} \begin{bmatrix} 1 & u_{12} & u_{13} & u_{14} \\ 0 & 1 & u_{23} & u_{24} \\ 0 & 0 & 1 & u_{34} \\ 0 & 0 & 0 & 1 \end{bmatrix} = \begin{bmatrix} 4 & 2 & -1 & 0 \\ 1 & -2 & 3 & 1 \\ 2 & -3 & 5 & 1 \\ -1 & 2 & -1 & 6 \end{bmatrix}$$

$l_{11} = 4$ $l_{11} = 4$

$l_{11} u_{12} = 2$ $u_{12} = 0 \cdot 5$

$$l_{11}u_{13} = -1 \qquad\qquad\qquad u_{13} = -0.25$$
$$l_{11}u_{14} = 0 \qquad\qquad\qquad\ u_{14} = 0$$

$$l_{21} = 1 \qquad\qquad\qquad\qquad l_{21} = 1$$
$$l_{21}u_{12} + l_{22} = -2 \qquad\qquad l_{22} = -2.5$$
$$l_{21}u_{13} + l_{22}u_{23} = 3 \qquad\quad u_{23} = -1.3$$
$$l_{21}u_{14} = l_{22}u_{24} = 1 \qquad\quad u_{24} = -0.4$$

$$l_{31} = 2 \qquad\qquad\qquad\qquad l_{31} = 2$$
$$l_{31}u_{12} + l_{32} = -3 \qquad\qquad l_{32} = -4$$
$$l_{31}u_{13} + l_{32}u_{23} + l_{33} = 5 \qquad l_{33} = 0.3$$
$$l_{31}u_{14} + l_{32}u_{24} + l_{33}u_{34} = 0 \qquad u_{34} = -2.0$$

$$l_{41} = -1 \qquad\qquad\qquad\qquad l_{41} = -1$$
$$l_{41}u_{12} + l_{42} = 2 \qquad\qquad\quad l_{42} = 2.5$$
$$l_{41}u_{13} + l_{42}u_{23} + l_{43} = -1 \qquad l_{43} = 2.0$$
$$l_{41}u_{14} + l_{42}u_{24} + l_{43}u_{34} + l_{44} = 6 \qquad l_{44} = 11.0$$

Hence, we solve

$$\begin{bmatrix} 4 & 0 & 0 & 0 \\ 1 & -2.5 & 0 & 0 \\ 2 & -4 & 0.3 & 0 \\ -1 & 2.5 & 2.0 & 11.0 \end{bmatrix} \begin{bmatrix} y_1 \\ y_2 \\ y_3 \\ y_4 \end{bmatrix} = \begin{bmatrix} 0 \\ -5 \\ -5 \\ 3 \end{bmatrix}$$

$$y_1 = 0, \qquad y_2 = 2, \qquad y_3 = 10, \qquad y_4 = -2$$

The upper-triangular form is then solved.

$$\begin{bmatrix} 1 & 0.5 & -0.25 & 0 \\ 0 & 1 & -1.3 & -0.4 \\ 0 & 0 & 1 & -2.0 \\ 0 & 0 & 0 & 1 \end{bmatrix} \begin{bmatrix} x_1 \\ x_2 \\ x_3 \\ x_4 \end{bmatrix} = \begin{bmatrix} 0 \\ 2 \\ 10 \\ -2 \end{bmatrix}$$

$$x_4 = -2, \qquad x_3 = 6, \qquad x_2 = 9, \qquad x_1 = -3$$

This can be checked in the original equations to show that the true solution has been obtained.

5 Solve the following tridiagonal set of equations.

$$\begin{bmatrix} 2 & 1 & 0 & 0 & 0 \\ 3 & -2 & -3 & 0 & 0 \\ 0 & 4 & -4 & 2 & 0 \\ 0 & 0 & -1 & 3 & -2 \\ 0 & 0 & 0 & 5 & 4 \end{bmatrix} \begin{bmatrix} x_1 \\ x_2 \\ x_3 \\ x_4 \\ x_5 \end{bmatrix} = \begin{bmatrix} 3 \\ 2 \\ 1 \\ 0 \\ -1 \end{bmatrix}$$

The Thomas algorithm given by Eqns (4.34), (4.35), and (4.36) is used

$$\alpha_1 = 2 \qquad\qquad\qquad\qquad \gamma_1 = \tfrac{1}{2}$$
$$\qquad\qquad\qquad\qquad\qquad\quad = 0.5$$
$$\alpha_2 = -2 - 3 \times 0.5 \qquad\qquad \gamma_2 = -3/(-3.5)$$
$$\quad = -3.5 \qquad\qquad\qquad\qquad = 0.857\,142$$
$$\alpha_3 = -4 - 4 \times 0.857\,143 \qquad \gamma_3 = 2/(-7.428\,572)$$
$$\quad = -7.428\,752 \qquad\qquad\qquad = -0.269\,231$$
$$\alpha_4 = 3 + 1 \times (-0.269\,231) \qquad \gamma_4 = 2/2.730\,769$$
$$\quad = 2.730\,769 \qquad\qquad\qquad\quad = -0.732\,394$$
$$\alpha_5 = 7.661\,970$$

$$u_1 = \tfrac{3}{2}$$
$$= 1{\cdot}5$$
$$u_2 = (2 - 3 \times 1{\cdot}5)/(-3{\cdot}5)$$
$$= 0{\cdot}714\,286$$
$$u_3 = (1 - 4 \times 0{\cdot}714\,826)/(-7{\cdot}428\,572)$$
$$= 0{\cdot}25$$
$$u_4 = (0 + 1 \times 0{\cdot}25)/2{\cdot}730\,769$$
$$= 0{\cdot}091\,549$$
$$u_5 = (-1 - 5 \times 0{\cdot}091\,549)/7{\cdot}661\,970$$
$$= -0{\cdot}190\,257$$

The solution of the back-substitution is

$$x_5 = -0{\cdot}190\,257$$
$$x_4 = 0{\cdot}091\,549 + 0{\cdot}732\,394 \times (-0{\cdot}190\,257) = -0{\cdot}047\,794$$
$$x_3 = 0{\cdot}25 + 0{\cdot}269\,231 \times (-0{\cdot}047\,794) = 0{\cdot}237\,132$$
$$x_2 = 0{\cdot}714\,286 - 0{\cdot}857\,143 \times 0{\cdot}237\,132 = 0{\cdot}511\,030$$
$$x_1 = 1{\cdot}5 - 0{\cdot}5 \times 0{\cdot}511\,030 = 1{\cdot}244\,485$$

These answers can be substituted in the original equations to check that the calculations are correct.

Problems

1 Calculate A^{-1} by triangular decomposition.

$$A = \begin{array}{ccc} 4 & 6 & 8 \\ 6 & 10 & 17 \\ 8 & 17 & 25 \end{array}$$

2 Solve the set of equations $AX = B$ where

$$A = \begin{pmatrix} 1 & 2 & 3 \\ 4 & 12 & 2 \\ 5 & 10 & 1 \end{pmatrix} \qquad B = \begin{pmatrix} 1 \\ 0 \\ -1 \end{pmatrix}$$

3 Use slide rule to solve the following equations by the Gauss–Seidel method.

$$9x_1 - 2x_2 = 4{\cdot}1$$
$$18x_2 - 2x_3 = 1{\cdot}3$$
$$2{\cdot}1x_1 - 15x_3 = 3{\cdot}2$$

4 Express A as a product $L \cdot L^T$.

$$A = \begin{bmatrix} 9 & -18 & 3 \\ -18 & 61 & 34 \\ 3 & 34 & 81 \end{bmatrix}$$

5 Factorize into upper- and lower-triangular matrices.

$$\begin{bmatrix} 1 & 7 & 5 \\ 4 & 29 & 24 \\ 10 & 64 & 29 \end{bmatrix}$$

6 Solve the following set of equations using the Gauss–Seidel method using as a first approximation $x_1 = 2\cdot0$, $x_2 = 0$, and $x_3 = 2\cdot0$.

$$3x_1 + 2x_2 - \tfrac{1}{2}x_3 = 6$$
$$x_1 - 4x_2 + x_3 = 4$$
$$-x_1 \qquad + 2x_3 = 2$$

Also, solve the equation by the Jacobi method and compare the rates of convergence.

5

Solution of ordinary differential equations

5.1 Introduction

The rate of change of a variable frequently occurs in physical problems and the mathematical equations associated with a problem are often formulated in terms of the derivatives of the variable. As a simple example, most readers will be familiar with the equations of motion relating the distance x, the velocity dx/dt, and the acceleration d^2x/dt^2. For some of these equations it is possible to obtain solutions by mathematical analysis but there are two problems which may prevent this.

Firstly, although some problems can be solved by fairly simple analysis there are many differential equations which require for their solution a high degree of mathematical training and skill. Secondly, a wider acquaintance with physical problems will soon show there are many differential equations for which the solutions cannot be represented in a simple mathematical form. Approximate methods of solution are then the only possible approach.

However, there are difficulties associated with the use of numerical methods and one should not too hastily resort to a numerical approach if mathematical analysis will yield an explicit formula for the solution. Even if the problem does not have an explicit solution it may be advantageous to perform some mathematical analysis in order to put the problem in a form more suitable for numerical computation.

It is well known that the process of integration introduces arbitrary constants which are determined by extra conditions given on the function or its derivatives; for example, a third-order differential equation would require three extra conditions to determine the three arbitrary constants which arise. As a trivial example, if we integrate $dy/dx = 2$ we obtain the solution $y = 2x + A$ and, if we were given the condition $y = 4$ when $x = 0$ then it follows that $A = 4$.

There are two types of problem which arise according to the manner in which these conditions are specified. If all the required conditions are given at a single point we have an initial-value problem, and the method of solution is to start at the known point and move step by step along the range of integration. However, if the conditions are given at more than one point there is not sufficient information to commence computation at any single point and the method of calculation involves either the solution of a set of simultaneous equations, or the use of estimated values at one point. These estimated values are then corrected by iteration as the calculation proceeds. This second type of problem is known as a boundary-value problem and there is considerable mathematical background which should be studied by those seriously concerned with this type of problem. The next few sections will be concerned only with the initial-value problem.

5.2 Initial-value problem

The general form for the initial-value problem is given by specifying the equation for the derivative and a condition on the function at one point only, i.e.,

$$\frac{dy}{dx} = f(x, y) \qquad (5.1a)$$

$$y(a) = s \qquad (5.1b)$$

Note, that $f(x, y)$ simply means some expression involving x and y, e.g., $\sin x + \sin y$, $x^2 + y^2$, etc.

At first sight these equations may seem too elementary to be of general use but it is possible to extend their scope considerably by a simple extension of the notation. Consider the problem

$$\frac{d^2y}{dx^2} + \sin x \cdot \frac{dy}{dx} + \cos x \cdot y = 3 \qquad (5.2a)$$

$$y(a) = 1, \qquad \left(\frac{dy}{dx}\right)_{x=a} = -1 \qquad (5.2b)$$

We let $z = dy/dx$ and the equations can then be put in the form

$$\frac{dz}{dx} = -\sin x \cdot z - \cos x \cdot y + 3$$

$$\frac{dy}{dx} = z \qquad (5.3a)$$

and

$$z(a) = -1$$
$$y(a) = +1 \qquad (5.3b)$$

See also Example 5.1. The equations are very similar to Eqns (5.1) but we now have two equations for derivatives and two for initial conditions. Those familiar with vector notation will see that if y, s, and f are replaced by vector quantities the same form of equations can be used. Thus, equations of any order which are linear in the various derivatives can be represented in the standard form of Eqn (5.1) and the methods of solution described below can be used.

The methods of numerical solution can be derived by various means including finite-difference formulae and truncated Taylor series. These derivations show that in each computational step an approximation is made which introduces an error. The usefulness of a method will be shown to depend not only on the size of these errors, but on the way in which they are magnified as we proceed along the range of integration. We consider the concept of consistency, which relates to the error introduced at a particular point, and the concept of stability which relates to the growth of error as the calculation proceeds.

However, before looking at the mathematics which lies behind the choice of a suitable formula it will be instructive to consider the solution of a first-order ordinary differential equation from a graphical point of view. Equation (5.1b) gives the initial value of the solution and Eqn (5.1a) gives the value of the derivative at any point. If we wish to follow the curve it seems reasonable to start at the known initial point (x_0, y_0), $x_0 = a$, $y_0 = s$, and move from there in the direction of the gradient at that point. As this straight line is followed we move away from the solution curve so after moving a small distance h in the x direction the new coordinates $(x_0 + h, y_1)$ are calculated and this point is used as a new base point. The gradient at the new point is calculated and a further step is made in this new direction to the point $(x_0 + 2h, y_2)$.

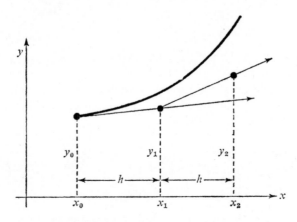

Fig. 5.1   Euler's method.

This process which is known as Euler's method is shown in Fig. 5.1. The procedure can be repeated step by step until the range of integration is covered. It should be noted that for convenience the step lengths were chosen to be equal although for this method it is possible to vary the size of the step h in different steps. It can be seen from the diagram that this method is unlikely to give a very accurate solution even if the step length is small since the approximate solution will always lie below and move away from a curve of the type shown. However, a much improved solution can be obtained by a simple modification.

After the first step has been taken an approximation to the solution y_1 is obtained at the point $x_0 + h$ and it is then possible to calculate the value of the derivative at this point. We now have values for the derivatives at both ends of the interval and it seems plausible that a better choice of direction from the initial point would be the average of these two values. A correction procedure is therefore used by starting from the initial point (x_0, y_0) and using the average gradient to calculate a new approximation to the solution for the point where $x = x_0 + h$ as shown in Fig. 5.2. This process of prediction followed by a single correction could then be continued to the end of the range. The calculation is summarized as follows.

1 Find $y_1^{(0)}$ as in the previous method.

$$y_1^{(0)} = y_0 + hf(x_0, y_0) \tag{5.4}$$

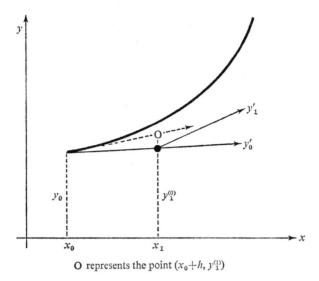

O represents the point $(x_0 + h, y_1^{(1)})$

Fig. 5.2 Euler–trapezoidal method.

2 Calculate the derivative at the point $(x_0 + h, y_1^{(0)})$.

3 Use the average of the derivatives at the ends of the intervals to progress from y_0 to a new approximation $y_1^{(1)}$.

$$y_1^{(1)} = y_0 + \frac{h}{2}[f(x_0, y_0) + f(x_0 + h, y_1^{(0)})] \tag{5.5}$$

4 Further corrections can be made at this stage by using the latest value to obtain a further correction.

$$y_1^{(r+1)} = y_0 + \frac{h}{2}[f(x_0, y_0) + f(x_0 + h, y_1^{(r)})], \qquad r = 1, 2, \ldots \tag{5.6}$$

5 After a sufficient number of corrections the new y value can be taken as the starting point of the next interval and steps 1–4 repeated with the subscripts incremented by one. This procedure is continued until the end of the interval is reached.

The above method, which is known as the Euler–trapezoidal method, is a simple example of a predictor–corrector formula in which an initial prediction is corrected by an iterative process. The iteration process shown above is a very primitive one and it should be noted that any standard iterative method could be used as an alternative.

The general form of the equation to be solved is given by

$$y_{n+1} = \phi_n + h\beta_{n+1}f(x_n + h, y_{n+1}) \tag{5.7}$$

The quantity ϕ_n represents those parts of the formula which depend on previous values and therefore do not change in the iteration. Unless the function $f(x, y)$ is such that Eqn (5.7) has an explicit solution an iterative solution is necessary and this can be obtained by writing the equation in the form

$$y_{n+1} - \phi_n - h\beta_{n+1}f(x_n + h, y_{n+1}) = 0 \tag{5.8}$$

and applying one of the methods of Chapter 2. The direct iteration procedure is most often employed but the method of Gear, which is discussed later in this chapter, employs a more sophisticated iteration process.

The methods used in practice are based on the same principle as the above method, but information from several previous x values is used to obtain greater accuracy. The other class of methods which is in widespread use is the Runge–Kutta type. These methods are based on taking estimates of the derivative not only at the ends of the interval, but at intermediate points also. The important feature of these methods is that the calculation uses only the values at the initial point and previous information is not used or stored.

5.3   Predictor–corrector methods

5.3.1   General form of the equations

The predictor–corrector methods in normal use are based on information from several previous points. This necessitates the storing of approximations to the solution and its derivative so that these can be used in later calculations. It also means that special attention must be given to the calculation of the first few values since past values are not available at that stage. However, we will first concentrate on the major problem of choosing suitable formulae.

The range of integration is divided into equal intervals h so that, for a solution in the interval $a \leqslant x \leqslant b$, we have

$$h = (b - a)/N \tag{5.9a}$$
$$x_n = a + nh, \qquad n = 0, 1, \ldots, N \tag{5.9b}$$

It is important to realize the distinction between the true solution, which is denoted by $y(x_n)$, and the approximation to the solution, which is denoted by y_n. In the first case we have a function defined for all values x ($a \leqslant x \leqslant b$) and in the second case we have a finite set of values y_n ($n = 0, 1, \ldots, N$). The notation f_n will be used to denote $f(x_n, y_n)$.

The traditional approach to the derivation of formulae is to look upon the problem as one of approximate integration and to use finite-difference methods to generate the formulae. If we integrate the differential equation (5.1a) between the limits x_n and x_{n+k} we have

$$y(x_{n+k}) - y(x_n) = \int_{x_n}^{x_{n+k}} f(x, y) \, dx \tag{5.10}$$

This integral cannot be evaluated directly since the solution y occurs under the integral sign. Various formulae are derived by the finite-difference calculus in texts such as Redish (1961) or *Modern Computing Methods* (1961) and this approach will not be further pursued here.

Formulae of this type, which are often used, are those based on the Newton–Cotes open or closed formulae which correspond to the predictor and corrector formulae respectively. Some simple examples of these formulae are given in Table 5.1. Other formulae to which we will refer frequently are the Adams–Bashforth and Adams–Moulton formulae, simple examples of which are included in Table 5.2.

It is interesting to derive the formulae from a more general viewpoint since this illustrates principles which are applicable to a variety of problems. The general method is to choose some formula containing a number of parameters which can be freely chosen by the user to satisfy some desirable mathematical or computational property. A set of conditions which the formula must satisfy can then be given, and these conditions determine acceptable values of the parameters. In this case the information available consists of the calculated values of the solution and its derivative at a sequence of points. The form chosen for the integration formulae is a linear combination of these

$$\sum_{r=0}^{k} \alpha_r y_{n+r} - h \sum_{r=0}^{k} \beta_r f_{n+r} = 0 \qquad (5.11)$$

where the α_r and β_r are constant coefficients which are chosen to give the best approximation. Since the size of the coefficients can vary by multiplying this equation by an arbitrary constant it will be convenient in the subsequent analysis to standardize the coefficients by dividing through by α_k which means that the coefficient α_k is always taken to be unity. Note that all the values y_{n+r} $(r = 0, 1, \ldots, k-1)$ are previously calculated values which have been stored and are immediately available. Therefore, any formula in which $\beta_k = 0$ enables y_{n+k} to be calculated explicitly. These formulae are variously known as explicit, open, or predictor formulae. In the case where $\beta_k \neq 0$ the quantity y_{n+k} also occurs in $f_{n+k} = f(x_{n+k}, y_{n+k})$ which is in general nonlinear and therefore, iteration will be required. These formula are known as implicit, closed, or corrector formulae. The coefficients, α_r and β_r, are chosen so that the requirements of accuracy and stability are satisfied.

5.3.2 Accuracy
One important property of a formula is the error introduced by the use of the approximation at a particular point. This will depend not only on the co-ordinates of the point considered but also on the derivative. Therefore, a general statement of which formula is more accurate than another is rather difficult. The way in which accuracy is often defined is to substitute the corresponding values of the true solution into the multistep formula (5.11) and expand each term as a Taylor series. The order of accuracy is then determined from the lowest power of h which has a nonzero coefficient. If the first non-vanishing term contains the power h^{p+1}, then the formula is said to have order of accuracy p. It is necessary that the order of accuracy be at least 1, and formulae which satisfy this requirement are said to be consistent. According to this definition of accuracy, the coefficients must be chosen so that p is as high as possible, i.e., as many as possible of the lower powers of h should have zero coefficients.

As an example, an explicit formula using the values at three points only will be considered and the coefficients will be chosen to give the most accurate formula. It will be shown, however, that the formula produced is computationally useless because it is strongly unstable; the errors grow in an uncontrollable fashion as the computation proceeds from step to step. The formula considered is

$$y_{n+2} + \alpha_1 y_{n+1} + \alpha_0 y_n - h[\beta_1 f_{n+1} + \beta_0 f_n] = 0 \qquad (5.12)$$

so that inserting the true solution values gives the formula for the error

$$y(x_{n+2}) + \alpha_1 y(x_{n+1}) + \alpha_0 y(x_n) - h[\beta_1 f(x_{n+1}, y(x_{n+1})) + \beta_0 f(x_n, y(x_n))] = E_{n+2}$$
(5.13)

The Taylor series expansions for the various terms are

$$y(x_{n+2}) = y(x_n) + 2hy'(x_n) + \frac{4h^2}{2}y''(x_n) + \frac{8h^3}{6}y'''(x_n) + \frac{16h^4}{24}y^{(IV)}(x_n) + \cdots \quad (5.14a)$$

$$y(x_{n+1}) = y(x_n) + hy'(x_n) + \frac{h^2}{2}y''(x_n) + \frac{h^3}{6}y'''(x_n) + \frac{h^4}{24}y^{(IV)}(x_n) + \cdots \quad (5.14b)$$

$$y(x_n) = y(x_n) \quad (5.14c)$$

$$f[x_{n+1}, y(x_{n+1})] = y'(x_{n+1}) = y'(x_n) + hy''(x_n) + \frac{h^2}{2}y'''(x_n) + \frac{h^3}{6}y^{(IV)}(x_n) + \cdots$$
(5.14d)

$$f[x_n, y(x_n)] = y'(x_n) \quad (5.14e)$$

The terms in α_r and β_r are now chosen so that the low-order terms in h have zero coefficients. Thus, the coefficient of $y(x_n)$ will be zero if $1 + \alpha_1 + \alpha_0 = 0$ etc. In this way four simultaneous equations are obtained which can be solved for the four unknowns.

$$1 + \alpha_1 + \alpha_0 = 0 \qquad 2 + \frac{\alpha_1}{2} - \beta_1 = 0$$

$$2 + \alpha_1 - \beta_1 - \beta_0 = 0 \qquad \frac{4}{3} + \frac{\alpha_1}{6} - \frac{\beta_1}{2} = 0$$
(5.15)

These equations have the solutions

$$\alpha_1 = 4 \qquad \alpha_0 = -5$$
$$\beta_1 = 4 \qquad \beta_0 = 2$$
(5.16)

and these coefficients give an error term of order h^4. (See also Example 5.5.)

5.3.3 Stability

The subject of stability is discussed in more detail in Section 5.5 but it is revealing to see that the pursuit of accuracy, as in the above example, can lead to disastrous results from a stability viewpoint. This will emphasize the fact that in numerical procedures one must always consider the way in which errors magnify through a process as well as the accuracy of each computational unit. For the sake of illustration we consider the simple equation

$$\frac{dy}{dx} = 0 \quad (5.17a)$$

$$y(0) = 1 \quad (5.17b)$$

which has the solution $y = 1$ for all x. While we would not use a numerical method for a simple equation such as this, we would certainly expect the method to be suitable for solving this equation.

The finite-difference equation (5.12) now has the form

$$y_{n+2} + 4y_{n+1} - 5y_n = 0 \quad (5.18)$$

which can easily be solved since it is a linear equation. A good treatment of the solution of finite-difference equations is available in Goldberg (1958) but it will suffice for our purpose to borrow some simple results. The solution of Eqn (5.18) can be found by substituting a trial solution $y_n = A(z)^n$ which leads to an auxiliary polynomial equation. The roots of the auxiliary polynomial, if distinct, give the values of z for which y_n satisfies the finite-difference equation. In this case the auxiliary equation resulting from the substitution is

$$Az^n[z^2 + 4z - 5] = 0 \qquad (5.19)$$

which has two meaningful solutions $z_1 = 1$ and $z_2 = -5$. The complete solution of the difference equation is

$$y_n = A_1(z_1)^n + A_2(z_2)^n$$
$$= A_1(+1)^n + A_2(-5)^n \qquad (5.20)$$

The values of A_1 and A_2 are found from the initial conditions and in this case, within the limits of accuracy of the computer, A_1 would be approximately unity and A_2 approximately zero. Thus, the first term corresponds to the solution but the second term is a spurious term which has most unsatisfactory properties. Such spurious terms always arise when multistep methods are used in an attempt to increase the accuracy of the scheme. In this case the magnitude of the spurious term is increased by a factor of 5 at each stage so that even if A_2 is small this term will eventually dominate the true solution. For example, after only 10 steps the term $(z_2)^n$ is approximately 10^7 and unless A_2 was initially less than 10^{-7}, the spurious term is already as large as the true solution and will be the most significant term in the solution from this point onwards.

It is clear that where formulae introduce spurious terms it is essential that these rapidly die away and conditions which will ensure this property are discussed in Section 5.5. The strong instability shown by the above equation can be easily removed by including the derivative term at the point x_{n+2} but not increasing the accuracy. This would give four equations in five unknowns and this gives one free parameter which can be chosen to give good stability properties as follows.

The set of equations now become

$$1 + \alpha_1 + \alpha_0 \qquad\qquad = 0 \qquad 2 + \frac{\alpha_1}{2} - 2\beta_2 - \beta_1 = 0$$

$$\qquad\qquad\qquad\qquad\qquad\qquad\qquad\qquad\qquad (5.21)$$

$$2 + \alpha_1 - \beta_2 - \beta_1 - \beta_0 = 0 \qquad \frac{4}{3} + \frac{\alpha_1}{6} - 2\beta_2 - \frac{\beta_1}{2} = 0$$

All the coefficients will be expressed in terms of a single coefficient, say α_0, and then the value of this coefficient will be chosen to give stability.

$$\alpha_1 = -1 - \alpha_0 \qquad \beta_2 = \frac{5}{12} + \frac{\alpha_0}{12}$$

$$\alpha_0 = \alpha_0 \qquad\qquad \beta_1 = \frac{2}{3} - \frac{2}{3}\alpha_0 \qquad\qquad\qquad (5.22)$$

$$\beta_0 = \frac{-1}{12} - \frac{5}{12}\alpha_0$$

By using the trial solution $A(z)^n$ as before we obtain the auxiliary equation

$$z^2 - (1 + \alpha_0)z + \alpha_0 = 0 \qquad (5.23)$$

which has solutions $z = 1$ or α_0.

In order to make sure the spurious solution does not increase as n increases we require that $-1 \leqslant \alpha_0 < +1$ giving several possible formulae which are stable, some examples of which are given below.

$$
\begin{array}{ccc}
\alpha_0 & -1 & 0 \\
\alpha_1 & 0 & -1 \\
\beta_0 & \frac{1}{3} & -\frac{1}{12} \\
\beta_1 & \frac{4}{3} & \frac{8}{12} \\
\beta_2 & \frac{1}{3} & \frac{5}{12}
\end{array}
\qquad (5.24)
$$

The first column is the well-known Simpson's rule and the second column is one of the implicit Adams–Moulton formulae which feature in the Nordsieck method in a modified form.

It should be noted that this approach was developed by studying the simple problem defined by Eqns (5.17). Any finite-difference scheme which is unstable for this example is not suitable for normal use. Formally, we say a scheme is strongly unstable if any root of the auxiliary equation has modulus greater than unity, or any multiple root has modulus equal to unity. However, even if we restrict our attention to schemes which are not strongly unstable, there are some schemes which are only weakly stable. For certain problems satisfactory results can be obtained but for other problems the methods should not be used. These terms are defined more formally, and further detailed discussion of this problem is given in Section 5.5.3. (See also Example 5.5.)

5.3.4　Method of computation

Although some time has been spent in showing the factors which influence the choice of coefficients the user will, in general, take the advice of some competent person on the method to be used and his concern will be to implement the method. The various steps involved are illustrated with regard to a multistep method using an Adams–Bashforth predictor and an Adams–Moulton corrector.

1 At the start of the process a method of finding the first few values is required since a typical multistep method uses values at several consecutive points and only one is available initially.

2 A predictor formula is used to obtain the first estimate of the new y value.

$$
y_{n+4}^{(0)} = y_{n+3} + \frac{h}{24}[55f_{n+3} - 59f_{n+2} + 37f_{n+1} - 9f_n] \qquad (5.25)
$$

3 The predicted value of the gradient is calculated.

$$
f_{n+4}^{(0)} = f(x_{n+4}, y_{n+4}^{(0)})
$$

4 A corrector formula is used iteratively.

$$
y_{n+4}^{(r+1)} = y_{n+3} + \frac{h}{24}[9f_{n+4}^{(r)} + 19f_{n+3} - 5f_{n+2} + f_{n+1}] \qquad r = 0, 1, \ldots
$$

$$
(5.26)
$$

5 After completing the iteration, steps 2–4 are repeated with all the subscripts increased by one to find the next y value and this process is repeated until the end of the interval of integration is reached.

The problem of finding the initial values for multistep schemes is solved in various ways in different schemes. The Taylor series expansion method could be used for a few steps, but this is more suitable for hand computation than computer use. (See Example 5.2.) Another alternative would be to use the single-step Euler–trapezoidal scheme with a step size smaller than that of the main integration scheme so that a comparable accuracy could be maintained. (See Example 5.4.) Alternatively a Runge–Kutta method as described in Section 5.4 is a suitable one-step scheme. In particular, the Runge–Kutta–Merson scheme would give some check that errors were small enough in the initial integration steps. The automatic methods of Nordsieck and Gear, described in Section 5.6, use a series of multistep methods of different orders, so that the process starts from a single-step scheme and builds up to the required order using a series of different formulae as further points become available. Some care is necessary in designing the starting procedure to ensure that the accuracy is at least as high as that of the main integration scheme. Since the main scheme normally involves a much higher power of h in the error term, this means that the starting methods should be based on a much reduced value of the step size h.

The question of how the corrector is used must now be considered. The simplest approach is to continue iteration for as many times as is necessary to ensure sufficient accuracy. Practically, this means continuing the iteration until the difference between successive iterates is less than a specified small quantity. It is, of course, necessary to have a convergent process and conditions which ensure convergence can easily be derived. The general form of the corrector equation is

$$y_{n+k} = h\beta_k f(x_{n+k}, y_{n+k}) + \phi_{n+k-1} \tag{5.27}$$

where ϕ_{n+k-1} includes all terms which involve the previously known values and which, therefore, do not change during the iteration process. The iteration equation is

$$y_{n+k}^{(r+1)} = h\beta_k f(x_{n+k}, y_{n+k}^{(r)}) + \phi_{n+k-1}, \qquad r = 0, 1, \ldots \tag{5.28}$$

and we are interested in the error given by

$$\begin{aligned} e_{n+k}^{(r+1)} &= y_{n+k} - y_{n+k}^{(r+1)} \\ &= h\beta_k[f(x_{n+k}, y_{n+k}) - f(x_{n+k}, y_{n+k}^{(r)})] \end{aligned} \tag{5.29}$$

Although more sophisticated conditions for convergence can be defined it is possible to give a simple condition which is sufficient for convergence if $f(x, y)$ is partially differentiable with respect to y. Using the mean-value theorem of differential calculus

$$f(x, y_1) - f(x, y_2) = (y_1 - y_2)\frac{\partial f}{\partial y}(x, \zeta) \tag{5.30}$$

where ζ lies in the interval (y_1, y_2). Let $K = \max |\partial f/\partial y|$ in the region of the iteration process. Then

$$\begin{aligned} |e_{n+k}^{(r+1)}| &= h|\beta_k|\left|\frac{\partial f}{\partial y}\right||y_{n+k} - y_{n+k}^{(r)}| \\ &\leqslant hK|\beta_k||e_{n+k}^{(r)}| \\ &\leqslant [(hK)|\beta_k|]^{r+1}|e_{n+k}^{(0)}| \end{aligned} \tag{5.31}$$

D

If $hK|\beta_k| < 1$ then the error decreases at each iteration and has limit zero as $r \to \infty$. The required condition for convergence is, therefore,

$$h\,|\beta_k|\left|\frac{\partial f}{\partial y}\right| < 1 \tag{5.32}$$

for the range of values of x and y occurring in the iteration sequence.

Thus, if we know the maximum value of $|\partial f/\partial y|$ a suitable size for h can be chosen. It should be noted that we are concerned with the value of $\partial f/\partial y$ not only in the range of the solution values but also over any values which might be reached in the iteration process. This latter range can, of course, be much larger than the solution range.

We must also consider whether the computation time involved in doing several iterations is worthwhile. In the case where the evaluation of the derivative takes most of the computing time (which is often the case) a doubling of the number of iterations would take the same time as doubling the number of steps by halving the step size h. The truncation error of a formula depends on a power of h, therefore, reducing the step size will decrease the likely error in a much stronger fashion.

If we consider the formula used in Eqn (5.25) which has an error term $(251/720)h^5 f^{IV}(\zeta_1)$ the effect of halving the interval will reduce the error by a factor of approximately 32 in each half-interval. Adding the two errors gives a factor of 16. However, the corrector formula has an error term $(-19/720)h^5 f^{IV}(\zeta_2)$ so that if the corrector were iterated to convergence the reduction in the error compared with the predictor formula would be a factor of the order $\frac{251}{720} \cdot \frac{720}{19} \approx 13$. Clearly, the step halving with a factor of 16 is preferable. Similar conclusions arise for most formulae so that iteration to convergence is in general uneconomic.

There is one refinement to the computational scheme using a predictor and corrector which has proved useful. A predictor and corrector are chosen with the same order of error term so that a linear combination of the predicted value and the corrected value can be used in an attempt to reduce the error. A motivation for this procedure is given below, but the reader should realize that a precise mathematical study of these problems is beyond the scope of this book. As an example of the method, the predictor equation (5.25) will be used together with the corrector equation (5.26). If these finite-difference equations are applied to the true solution this gives

$$y(x_{n+4}) = y(x_{n+3}) + \frac{h}{24}[55f(x_{n+3}, y(x_{n+3})) - 59f(x_{n+2}, y(x_{n+2}))$$

$$+ 37f(x_{n+1}, y(x_{n+1})) - 9f(x_n, y(x_n))] + \frac{251}{720}h^5 f^{IV}(\zeta_1, y(\zeta_1)) \tag{5.33}$$

and

$$y(x_{n+4}) = y(x_{n+3}) + \frac{h}{24}[9f(x_{n+4}, y(x_{n+4})) + 19f(x_{n+3}, y(x_{n+3}))$$

$$- 5f(x_{n+2}, y(x_{n+2})) + f(x_{n+1}, y(x_{n+1}))] - \frac{19}{720}h^5 f^{IV}(\zeta_2, y(\zeta_2)) \tag{5.34}$$

where ζ_1, ζ_2 are values of x within the range $[x_n, x_{n+4}]$

The method which is used depends upon two assumptions and significant departure from these may well make the method unsatisfactory. Firstly, it is assumed that $f^{IV}(\zeta, y(\zeta))$ is constant in the interval so that the error terms differ only by the constant multiplier. The error involved in this assumption for a well-behaved function will be small over a short interval. Also, the first five terms on the right-hand side are replaced by the computed values at corresponding points when the value of y_{n+4} is calculated. Thus, the error term of the two equations above will only represent the actual error if the values used in the finite-difference equation are fortuitously correct.

The basis of the method is to use the two equations to eliminate the error represented by the h^5 term but the reader should bear in mind both the presence of accumulated error and the approximate nature of the assumptions. As an example, if the predicted and corrected values are respectively $y_{n+4} = 3\cdot195$ and $y_{n+4} = 3\cdot141$ then

$$3\cdot195 = y_{n+3} + \frac{h}{24}[55f_{n+3} - 59f_{n+2} + 37f_{n+1} - 9f_n]$$

$$(5.35)$$

$$3\cdot141 = y_{n+3} + \frac{h}{24}[9f_{n+4} + 19f_{n+3} - 5f_{n+2} + f_{n+1}]$$

Equations (5.33) and (5.34) now become

$$y(x_{n+4}) \approx 3\cdot195 + \tfrac{251}{720}E$$
$$y(x_{n+4}) \approx 3\cdot141 - \tfrac{19}{720}E \qquad (5.36)$$

It is now possible to eliminate E

$$(19 + 251)y(x_{n+4}) = 19 \times 3\cdot195 + 251 \times 3\cdot141 \qquad (5.37)$$
$$y(x_{n+4}) = 3\cdot145$$

At each step the process then follows the three stages of prediction, correction, and improvement before moving to the next step.

5.4 Runge–Kutta methods

5.4.1 Some typical Runge–Kutta formulae

The previous methods used the values of the function and its derivatives at several previous points. This causes difficulty both in starting the method and in reducing the interval since in these cases the past values are not available. The Runge–Kutta methods generate all the function values and derivatives which are required by starting from one initial point in each computational step. Thus, they are ideal for starting the solution or for changing the step size, but have the disadvantage that several evaluations of the derivative per step are required. The derivation of the Runge–Kutta formula will not be shown since this is rather involved, but a typical example will be used to show both the motivation behind the scheme and the computational procedure.

A popular fourth-order scheme is given by the following equations where for simplicity a constant step size is shown. (See Example 5.6.) In a more detailed scheme the step size would be varied as the integration proceeded in order to meet some accuracy criteria.

$$k_1 = hf(x_n, y_n)$$
$$k_2 = hf(x_n + h/2, y_n + k_1/2)$$
$$k_3 = hf(x_n + h/2, y_n + k_2/2) \tag{5.38}$$
$$k_4 = hf(x_n + h, y_n + k_3)$$
$$y_{n+1} = y_n + \tfrac{1}{6}(k_1 + 2k_2 + 2k_3 + k_4) \tag{5.39}$$

Since, $dy/dx = f(x, y) \approx \Delta y/h$ it can be seen that the k_r represent estimates of the increment in y at the left-end point, twice at the midpoint, and once at the right-end point. At each calculation all the previous estimates of the increment k_r are available and can be used to calculate the value of y in $f(x, y)$. A weighted average value of the y increments is then added to the previous y value to obtain the next solution point.

In the above formula only one k_r appears in each right-hand side which makes it a popular formula for desk-machine use, but since the k_r can be used in different combinations on the right-hand side several formulae are possible of the general form

$$k_r = h_n f(x_n + \alpha_r h_n, y_n + \sum_{s=1}^{r-1} \beta_{rs} k_s) \tag{5.40}$$

Note, that a subscript n has been used with the step size to remind the reader that the step size can be varied at each step if required.

One of the main difficulties in using the Runge–Kutta formulae is due to the fact that the error term is very complicated and cannot easily be used for estimation of the truncation error or for determining a suitable step size. However, a method has been designed which uses a simple approximation to give an estimate of the error which can be used to vary the step size. This method, which is known as the Runge–Kutta–Merson method, uses one extra derivative calculation and provides an error estimate which is exact when the derivative function is linear in x and y. Hopefully, if the step size is small enough the linearity of the derivative will be a reasonable approximation.

The equations for the Runge–Kutta–Merson method are

$$k_1 = hf(x_n, y_n)$$
$$k_2 = hf(x_n + h/3, y_n + k_1/3)$$
$$k_3 = hf(x_n + h/3, y_n + k_1/6 + k_2/6) \tag{5.41}$$
$$k_4 = hf(x_n + h/2, y_n + k_1/8 + 3k_3/8)$$
$$k_5 = hf(x_n + h, y_n + k_1/2 - 3k_3/2 + 2k_4)$$
$$y_{n+1} = y_n + \tfrac{1}{6}[k_1 + 4k_4 + k_5] \tag{5.42}$$

It should be noted that both this formula and the previous equation (5.38) have an error term of order h^5 but in this case one extra derivative must be calculated. The error estimate is given by

$$E \approx \tfrac{1}{30}[2k_1 - 9k_3 + 8k_4 - k_5] \tag{5.43}$$

and this estimate is exact if $f(x, y) = Ax + By + C$.

The accuracy of a Runge–Kutta formula is usually defined in terms of Taylor series expansion, as in the case of predictor–corrector formulae, but in this case an expansion in terms of the two variables x and y is required. The order of accuracy is again determined from the first non-vanishing power of h in the expansion. The details of these calculations are available in Conte (1965) and Bull (1966).

5.4.2 Stability

The stability properties of Runge–Kutta formulae are rather different from those of the multistep methods since the Runge–Kutta methods do not introduce spurious solutions. In the case of multistep methods these spurious solutions must contribute a negligible amount to the solution, but the problem with Runge–Kutta methods is simply the accuracy with which the finite-difference solution approximates the true solution. The instability is known as partial instability because a reduction in the size of h will eliminate the instability; it may however be uneconomic to reduce h to the size required. The situation in which this instability is most obvious is when the true solution is a decreasing exponential function. This can be converted to an increasing finite-difference solution by choosing too large a size of h. Normally, the restriction on h which is given takes the form

$$h\left|\frac{\partial f}{\partial y}\right| < K$$

where K takes a value between 2·0 and 2·8. This instability is a problem when a dominant component would allow a large step which would change a decreasing exponential component into a rapidly increasing finite difference solution. The instability is also of some concern when a set of equations are to be solved since the partial stability criterion must be satisfied for each component, which can necessitate a very small step size. It should be noted that the Runge–Kutta–Merson method will give a step size even smaller than that defined by the partial-stability criteria above. This is discussed in more detail in Section 5.5.4.

5.4.3 Extension to simultaneous differential equations

The extension of Runge–Kutta methods to simultaneous equations is quite simple and the relevant equations are shown here. If we consider the problem

$$y' = f(x, y, z) \qquad z' = g(x, y, z)$$
$$y_0 = s_1 \qquad\qquad z_0 = s_2 \tag{5.44}$$

then the following computational scheme can be used

$$k_1 = hf(x_n, y_n, z_n) \qquad m_1 = hg(x_n, y_n, z_n)$$
$$k_2 = hf(x_n + h/2, y_n + k_1/2, z_n + m_1/2) \tag{5.45}$$
$$m_2 = hg(x_n + h/2, y_n + k_1/2, z_n + m_1/2)$$

and so on.

$$y_{n+1} = y_n + \tfrac{1}{6}[k_1 + 2k_2 + 2k_3 + k_4] \tag{5.46a}$$
$$z_{n+1} = z_n + \tfrac{1}{6}[m_1 + 2m_2 + 2m_3 + m_4] \tag{5.46b}$$

The only point to note is that k_1 and m_1 must both be calculated before k_2 and m_2. The extension of the above scheme to several variables is trivial and easy to program on a computer.

5.5 Stability

5.5.1 Inherent and induced instability

When computer calculations are of a step-by-step character, it is important to consider not only the size of the errors introduced at each step, but also the way in which these errors grow throughout the process. The study of stability is concerned with this latter property. Various authors use the term stability in rather different ways so that some care is necessary in understanding the meaning of the term.

There are two basic types of instability to be distinguished, inherent instability and induced instability. In the case of inherent instability small variations in the conditions of the problem cause large variations in the true solution and this instability must inevitably be reflected in the finite-difference scheme. Induced instability refers to error magnification which is introduced by the method of solution and may not be serious if the error growth is no worse than the growth of the true solution.

It is important to distinguish between instability of the multistep methods which occurs when one of the spurious solutions grows rapidly to dominate the true solution, and the partial instability of the Runge–Kutta methods where the finite-difference solution represents the true solution inadequately because too large a step size is used.

Stability concepts must be formulated very carefully in the case where the true solution is increasing. It is quite likely in this case that the errors will also grow and the property which is required here is that the errors grow no faster than the true solution. A scheme will be called relatively stable if this property holds so that, although the absolute error is magnified, the relative error does not grow. As yet, standard procedures which will test for relative stability are not available so that this is more an important theoretical concept than a practical tool.

5.5.2 Inherent instability

This occurs when small variations in the conditions which define the differential problem cause a large variation in the solution of the differential equation and the instability is a fundamental feature of the problem formulation. Since the finite-difference scheme follows the properties of the differential equation as closely as possible, it will also magnify small perturbations which will inevitably be present in the form of round-off or truncation error. This difficulty is inherent in the formulation of the problem and the only way to remove the instability is to try and reformulate the problem so that the strong sensitivity to perturbations disappears.

An example of this instability occurs when the differential solution contains a dominant term which is eliminated by a particular set of initial conditions. Consider the equation

$$y' = y - x \tag{5.47}$$

which has the exact solution

$$y = Ae^x + x + 1 \tag{5.48}$$

If the initial conditions are $y = 1$ when $x = 0$ then we have $A = 0$ and the dominant part of the solution e^x is theoretically absent. However, small errors in the finite difference scheme will introduce this component and owing to its rapid growth it will soon swamp the true solution.

5.5.3 Strong instability and weak stability

There are special problems associated with multistep methods since the use of the more accurate higher-order methods automatically introduces extra solutions and the character of these spurious solutions is crucial. Ideally, they should rapidly die away or, at the very least, not increase.

The most catastrophic situation is where one of the spurious solutions increases in a manner which cannot be rectified by reducing the size of h; in fact, the solution in a strongly unstable example grows without bound as h tends to zero. An example of a strongly unstable scheme was presented in Section 5.3.2. Such schemes are unacceptable but it may also be true that schemes which are not strongly unstable are unsuitable for some classes of problem. The weakly stable schemes have increasing spurious solutions for some problems and care must be exercised when using weakly stable methods.

These properties are defined mathematically as follows. We consider the differential equation

$$y' = f(x, y)$$
$$y(a) = s \tag{5.49}$$

and use the following finite-difference equation

$$\sum_{r=0}^{k} \alpha_r y_{n+r} - h \sum_{r=0}^{k} \beta_r f_{n+r} = 0 \tag{5.50}$$

If $k > 1$ then the difference scheme has more solutions than the differential scheme and the instabilities associated with multistep methods may arise. The analysis of these methods has been presented with full mathematical rigour by Dahlquist (1956) and Henrici (1962). It will suffice here to present the results of their investigations without proof.

The difference equation (5.50) which is to be solved for y_{n+k} is a non-linear equation with constant coefficients. For convenience of notation the polynomial equations

$$p(z) = \sum_{r=0}^{k} \alpha_r z^r \tag{5.51}$$

and

$$\sigma(z) = \sum_{r=0}^{k} \beta_r z^r \tag{5.52}$$

are introduced. If we insist that all difference schemes should be stable for the case where y is constant, i.e., $f(x, y) \equiv 0$, then only the linear portion of the difference scheme is left, and the solutions are given by the roots of the auxiliary equation $p(z) = 0$. If the roots of this equation are z_r $(r = 1, 2, \ldots, k)$ then the complete solution if the roots are all different is

$$y_n = A_1(z_1)^n + \cdots + A_k(z_k)^n \tag{5.53}$$

There always exists a principal root $z_1 = 1$ which corresponds to the true solution and strong instability arises when any z_r $(r \neq 1)$ has modulus greater than unity.

It can be seen that such a root increases without bound as n increases and makes the scheme useless. It can also be shown that this unbounded growth occurs if there are repeated roots z_r with $|z_r| = 1$.

If all the roots satisfy the condition $|z_r| \leqslant 1$, and multiple roots the condition $|z_r| < 1$, then strong instability cannot arise, but a method may exhibit the property known as weak stability. This difficulty occurs in certain finite-difference schemes but only for particular classes of problems. A weakly stable scheme may be quite satisfactory for certain derivative functions but give poor results in other problems. This arises because the complete finite-difference equation will have roots which are different from those given by the auxiliary equation since the terms $h\beta_r f_{n+r}$ must also be considered. It can happen that one or more of the roots of this full equation lie outside the unit circle and lead to an increasing spurious solution which is the cause of weak stability.

This behaviour is particularly evident when the true solution is a strongly decreasing one such as the solution of the following equation

$$y' = -y$$
$$y(0) = 1 \tag{5.54}$$

which has the true solution $y = e^{-x}$. For example, the Simpson corrector

$$y_{n+2} = y_n + \frac{h}{3}[f_{n+2} + 4f_{n+1} + f_n] \tag{5.55}$$

has the solution

$$y_n = A_1[z_1]^n + A_2[z_2]^n \tag{5.56}$$

and the z_r are approximations to e^{-h} and $-e^{h/3}$. If the starting values are accurate this will ensure that A_1 is approximately unity and A_2 approximately zero so that initially the second component will be small. However, since this component is increasing exponentially it will eventually dominate the true solution. The size of this second component depends on the size of A_2 and on the product hn. Thus, the effect of the weak stability can be reduced in two ways. If the starting conditions are made very accurate, for example by reducing the step length h, then A_2 will be small. If the length of the interval of integration is small then hn is also small so that $e^{hn/3}$ remains small. It is clear that weakly stable schemes are unsuitable for integration over long intervals.

Dahlquist proved several important properties of multistep schemes including the following:

1 The order of accuracy of a multistep method must be restricted if the scheme is to be stable. Since there are $2k + 2$ coefficients in Eqn (5.11) it is theoretically possible to achieve an accuracy of order $2k + 1$. However, the order of a stable operator where k is odd cannot exceed $k + 1$. If k is even, the order of a stable operator cannot exceed $k + 2$.

2 The necessary and sufficient conditions to achieve the maximum order of accuracy $k + 2$, while maintaining stability, is that k is even, all the roots of $p(z)$ are of modulus unity, and the following relations hold between the coefficients of $p(z)$ and $\sigma(z)$.

$$\alpha_r = -\alpha_{k-r}, \qquad \beta_r = \beta_{k-r} \tag{5.57}$$

3 A stable formula with maximum order of accuracy will always have the property of weak stability.

5.5.4 Partial stability

Partial instability is the term used to describe the phenomenon where too large a step size h causes a decreasing solution to be represented by an increasing finite-difference solution. The term is frequently used in connection with Runge–Kutta methods, and the instability is particularly evident when these methods are used for the solution of stiff equations, since it then happens that the most demanding condition on h is given by the least significant component of the solution. The term stiff equation is used when components of a solution are decreasing at widely differing rates, for example, in Eqns (5.63)–(5.66). It is not a true instability since it results from an inaccurate representation of one of the solution components rather than a magnification of errors.

The following examples illustrate how this type of instability arises. We consider the second order Runge–Kutta method

$$k_1 = hf(x_n, y_n)$$
$$k_2 = hf(x_n + h, y_n + k_1)$$
$$\quad (5.58)$$
$$y_{n+1} = y_n + \tfrac{1}{2}[k_1 + k_2] \quad (5.59)$$

applied to the first-order ordinary differential equation

$$y' = \lambda y$$
$$y(0) = A$$
$$\quad (5.60)$$

which has the solution $y = Ae^{\lambda x}$. The Runge–Kutta formula gives

$$y_{n+1} = y_n\left[1 + h\lambda + \frac{h^2\lambda^2}{2}\right] \equiv y_n\phi_n \quad (5.61)$$

and the graph in Fig. 5.3 shows how this multiplier varies for $\lambda = \pm 1$.

For the case $\lambda = 1$ we see the finite-difference scheme increases the solution at each step in order to follow the true solution and although this means that error will also be magnified, this cannot be classed as a true instability since it is inherent in the differential equation. In such a case one requires that the scheme is relatively stable, i.e., that the errors in the difference scheme grow no faster than the solution, but this is a difficult topic which will not be pursued further.

For the case $\lambda = -1$ a scheme which was relatively stable would need to decrease at the same rate as e^{-x}, and we can see that this is only the case over a small range of $h\lambda$. Indeed, for $h\lambda > 2$ the errors will actually be magnified and this phenomenon has been given the name partial instability. However, where the exponential component is a significant part of the solution, the condition $h\lambda < 2$ will not be stringent enough to give adequate accuracy as well as stability.

The above problem has received particular attention in the case of sets of equations with decreasing exponential solutions with widely differing values of λ. These have become known as stiff equations. In this case it is necessary for the condition

$$|h\lambda_i| < 2 \quad (5.62)$$

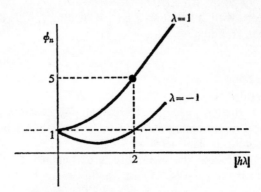

Fig. 5.3   Graph of Runge–Kutta factor ϕ_n

to apply for all the λ_i otherwise one of the decreasing solutions will be approximated by an increasing finite-difference solution which will soon dominate the other solutions. Runge–Kutta methods are not suitable for this type of problem and some method such as that of Gear should be used. This method is discussed in Section 5.6.

A simple stiff equation will illustrate this problem. The differential equation

$$y'' + 101y' + 100y = 0 \tag{5.63}$$

can be represented by the set of equations

$$\begin{aligned} y' &= z \\ z' &= -(100y + 101z) \end{aligned} \tag{5.64}$$

where y' denotes the derivative with respect to x. If the initial conditions are

$$\begin{aligned} y(0) &= 2 \\ z(0) &= -101 \end{aligned} \tag{5.65}$$

then the true solution is given by

$$y = e^{-x} + e^{-100x} \tag{5.66}$$

When $x = 1$ the first part of the solution $\approx 0\cdot368$ and the second part is less than 10^{-8}. The temptation is to say that the second part of the solution is negligible and can, therefore, be ignored. However, if the partial-stability condition is ignored then the second part of the solution will start to grow and will soon be larger than the first part. It is, therefore, essential that condition (5.62) holds for all values of λ_i and this can give a severe limitation on the size of h.

5.6   Further methods

In view of the extra computation which results from using Runge–Kutta methods rather than predictor–corrector methods, it would be useful if a predictor–corrector type method were available with built-in procedures to deal with the two problems of starting the solution and varying the step size. Nordsieck (1962) describes a computer program which has these features and uses two tests to adjust the step size, one to control the truncation error and the other to ensure that the method is stable. The method is essentially a reformulation of the Adams type finite-difference formulae which have very good stability properties for small $h\lambda$. The terms which are stored are those of the Taylor series expansion. The advantage of this is that where large derivatives are involved an element of scaling takes place owing to the presence of a power of h in the quantity stored. The interested reader should refer to the paper by Nordsieck (1962) for further details.

Although the details of the method may not be of interest to everyone the method may be available in libraries of computer programs and is recommended as a general-purpose program rather than Runge–Kutta–Merson which is more time-consuming.

However, there are the problems involving stiff equations, for which the above method is not ideal, owing to the very small step size which would be chosen by the automatic procedure. A method due to Gear (1968) is available which gives a dramatic improvement in the amount of time required for the solution of stiff equations and incorporates an automatic starting procedure. The method also incorporates a more sophisticated iterative solution of the corrector equation but this does not affect the basis of the method for stiff equations.

It is interesting to study the motivation behind the method of Gear since, although such a discussion is not essential for the user of numerical methods, it provides a useful insight into the problems of choosing an optimum method for non-stiff equations. Consider the general multistep equation

$$\sum_{r=0}^{k} \alpha_r y_{n+r} - h \sum_{r=0}^{k} \beta_r f_{n+r} = 0 \tag{5.67}$$

and the corresponding equation obtained by substituting the solution values $y(x_{n+r})$ for y_{n+r} and $f(x_{n+r}, y(x_{n+r}))$ for f_{n+r}. This equation gives the error term, i.e.,

$$\sum_{r=0}^{k} \alpha_r y(x_{n+r}) - h \sum_{r=0}^{k} \beta_r f(x_{n+r}, y(x_{n+r})) = E_{n+k} \tag{5.68}$$

Subtract the first equation from the second and use the mean-value theorem.

$$f(x_{n+r}, y(x_{n+r})) - f(x_{n+r}, y_{n+r}) = [y(x_{n+r}) - y_{n+r}]\frac{\partial f}{\partial y}(x_{n+r}, \zeta) \tag{5.69}$$

where ζ lies in the interval given by the end points $y(x_{n+r}), y_{n+r}$. This gives

$$\sum_{r=0}^{k} \alpha_r e_{n+r} - h \sum_{r=0}^{k} \beta_r \lambda(x_{n+r}, \zeta) e_{n+r} = E \tag{5.70}$$

where

$$e_{n+r} = y(x_{n+r}) - y_{n+r} \quad \text{and} \quad \lambda \equiv \frac{\partial f}{\partial y}$$

The study of stability, in the sense of Dahlquist, is concerned with the roots of the auxiliary polynomial associated with the difference equation on the left-hand side of Eqn (5.70), namely

$$\sum_{r=0}^{k} \alpha_r z^r - h \sum_{r=0}^{k} \beta_r \lambda(x_{n+r}, \zeta) z^r = 0 \tag{5.71}$$

There is a root of this equation $z = 1$ when $h = 0$ but apart from this root the stability requirement is that all other roots should have modulus less than unity. Hopefully, we also require this condition for as high a value of h as possible.

There are two extremes which can be considered according to whether $h\lambda$ is negligible or very large. If $h\lambda$ is very small then the roots are small perturbations of the roots of the equation $\rho(z) \equiv \sum_{r=0}^{k} \alpha_r z^r = 0$, and it seems reasonable to make all roots equal to zero other than the principal root $z = 1$. The hope is that as $h\lambda$ grows, the roots will not quickly grow greater than unity, since they start from zero when $h = 0$. This gives $\rho(z) = z^k - z^{k-1}$ which is the equation associated with the method of Nordsieck. Now, since h will normally be chosen as large as possible, subject to stability and accuracy constraints, it follows that when $h\lambda$ is very small then λ must also be very small, and in the case of several variables all the λ_i must be very small.

In the case of stiff equations however, it is known that at least one λ_i is large and it is worth considering Eqn (5.67) when $h\lambda_i$ is large. If λ_i were constant the roots would be small perturbations of the roots of

$$\sigma(z) \equiv \sum_{r=0}^{k} \beta_r z^r = 0 \tag{5.72}$$

and it seems reasonable that the equation chosen should be z^k so that all the roots are equal to zero. There must also be a principal root $z_1 = 1$ of the polynomial $\rho(z)$ and these conditions define the equations used by Gear. The method of Gear is based on this auxiliary equation and it is, therefore, plausible that good results would be obtained for stiff equations, and this is certainly confirmed by experimental results.

It should be borne in mind that intuitive reasoning such as that above gives a useful insight into the background of the methods, but firm conclusions can only be based on rigorous mathematical analysis. In particular it should be clear that for general values of $h\lambda$ neither $\rho(z)$ nor $\sigma(z)$ dominate and the choice of the most stable method is by no means clear.

5.7   Comparison of methods

The Runge–Kutta methods are often used for the solution of engineering problems on a computer because they need no special starting method and are therefore very easy to program. Also, the Runge–Kutta–Merson method gives an automatic procedure for adjusting the step size, which again makes for easy use by the non-specialist. However, there are three disadvantages of the Runge–Kutta methods which cannot be overcome, whereas the two features noted above can be introduced into multistep methods with some careful programming, such as in the methods of Gear and Nordsieck. These conditions mean that multistep methods are generally preferable except where simplicity is the most important feature.

An important disadvantage of the Runge–Kutta methods is that the form of the error term is extremely complicated, and this is not completely overcome by the Merson method which is approximate only. Any analysis of the methods to find bounds on the errors is very difficult, whereas a simple formula for the error exists in multistep methods, provided the appropriate derivative can be calculated. Another undesirable feature of Runge–Kutta methods is the large number of derivative evaluations required per step. The Runge–Kutta–Merson method requires five such evaluations, whereas in a predictor–corrector method it is often found that two or three corrections are sufficient. This saving in computer time in the case of multistep methods can be very significant in the case where the derivative function involves the calculation of special functions or is complicated to evaluate.

The third problem associated with the use of Runge–Kutta methods has been mentioned in the discussion on stability where it was shown how inadequate the Runge–Kutta methods are for dealing with stiff equations. By contrast the method of Gear has shown a considerable saving in time over previous methods. A choice between the methods of Gear and Nordsieck will depend upon a knowledge of the values of λ_i and the eigenvalues of the jacobian $\partial f/\partial y$ for a system of equations. In the case where any of the λ_i is large the method of Gear is recommended, whereas the method of Nordsieck would be preferable if all the λ_i were small. The above recommendations are tentative only since complete analysis of a particular problem would be necessary before a rigorous conclusion could be reached.

Table 5.1

Newton–Cotes methods

Open

$y_{n+1} = y_{n-1} + 2hy'_n$ Midpoint

$y_{n+1} = y_{n-2} + \dfrac{3h}{2}[y'_n + y'_{n-1}]$

$y_{n+1} = y_{n-3} + \dfrac{4h}{3}[2y'_n - y'_{n-1} + 2y'_{n-2}]$

Closed

$y_{n+1} = y_n + \dfrac{h}{2}[y'_{n+1} + y'_n]$ trapezoidal

$y_{n+1} = y_{n-1} + \dfrac{h}{3}[y'_{n+1} + 4y'_n + y'_{n-1}]$ Simpson

$y_{n+1} = y_{n-2} + \dfrac{3h}{8}[y'_{n+1} + 3y'_n + 3y'_{n-1} + y'_{n-2}]$

Table 5.2
The open predictor-type methods are based on integration of Newton's backward-difference formula. The closed corrector-type equations are derived in the same way with a different value of p.

Adams–Bashforth methods

Open

$\nabla y_{n+1} = h(1 + \tfrac{1}{2}\nabla + \tfrac{5}{12}\nabla^2 + \cdots)y'_n$

$y_{n+1} = y_n + hy'_n$ Euler

$y_{n+1} = y_n + \dfrac{h}{2}[3y'_n - y'_{n-1}]$

$y_{n+1} = y_n + \dfrac{h}{12}[23y'_n - 16y'_{n-1} + 5y'_{n-2}]$

Closed

$$\nabla y_{n+1} = h(1 - \tfrac{1}{2}\nabla - \tfrac{1}{12}\nabla^2 - \cdots)y'_{n+1}$$

$$y_{n+1} = y_n + hy'_{n+1}$$

$$y_{n+1} = y_n + \frac{h}{2}[y'_{n+1} + y'_n] \quad \text{trapezoidal}$$

$$y_{n+1} = y_n + \frac{h}{12}[5y'_{n+1} + 8y'_n - y'_{n-1}]$$

Bibliographic notes

There are a large number of texts which give elementary treatments of the solution of ordinary differential equations. McCracken and Dorn (1964) and Conte (1965) contain a simple treatment including consideration of the error analysis and programming associated with standard methods. The books by Redish (1961), Bull (1966), and *Modern Computing Methods* (1961) give a coverage of the methods based on finite differences and the first two have several examples which are suitable for computation on a desk machine. A rather deeper treatment is given in Fox and Mayers (1968) with consideration of a wider range of methods than the simplest ones. The book is directed to the user rather than the theoretician and mathematical formulation with theorems and proofs is avoided. The book contains ample discussion of the problems of instability and several numerical examples as illustrations. The book by Henrici (1964) gives a formal treatment in mathematical style of some simpler methods which is suitable for undergraduate courses for mathematicians. Two other works are available by Henrici (1962) and (1963) which give an extensive treatment of predictor–corrector methods, Runge–Kutta methods, stability and error analysis, and associated topics. These are research texts suitable for the specialist in this field, but are not recommended for the normal user of computational methods. The texts by Ralston (1965) and Hamming (1962) provide a treatment which includes mathematical detail, but is still directed towards the user of numerical methods. A discussion which concentrates on the mathematical analysis involved in the numerical solution of ordinary differential equations is given in Isaacson and Keller (1966).

The boundary-value problem has not been considered here but the reader interested in this topic will find details in Conte (1965) and Fox and Mayers (1968). Mathematical analysis associated with this topic is given in Isaacson and Keller (1966) and Keller (1968).

Worked examples

1 Transform the following equation into the standard form of Eqn (5.1).

$$a(x)y'' + b(x)y' + c(x)y = 0, \qquad y(0) = s_1, \qquad y'(0) = s_2$$

Let $y' = z$ and $y'' = z$ and assume $a(x) > 0$.

$$z' = -\frac{b(x)}{a(x)}z - \frac{c(x)}{a(x)}y$$

Hence,

$$\begin{pmatrix} y \\ z \end{pmatrix}' = \begin{pmatrix} 0 & 1 \\ -\dfrac{c(x)}{a(x)} & -\dfrac{b(x)}{a(x)} \end{pmatrix}\begin{pmatrix} y \\ z \end{pmatrix}$$

$$\begin{pmatrix} y \\ z \end{pmatrix}_{x=0} = \begin{pmatrix} s_1 \\ s_2 \end{pmatrix}$$

This is now solved as two simultaneous first-order equations. Any higher-order equation which is linear in the various derivatives can be represented in this vector form.

2 Find the value of y at the points $x = 1 \cdot 0, 2 \cdot 0, 0 \cdot 5$ by the method of Taylor series for the equation $dy/dx = e^x$.

$$y(0) = 1$$

First find expressions for the various derivatives.

$$\frac{d^2y}{dx^2} = e^x, \quad \frac{d^3y}{dx^3} = e^x, \quad \text{etc.}$$

Starting at $x = 0$ we have $y(0) = 1$, $y'(0) = 1$, $y''(0) = 1$, etc. Using Taylor series

$$y(x + h) = y(x) + hy'(x) + \frac{h^2 y''}{2!}(x) + \cdots + \frac{+h^{n-1}}{(n-1)!} y^{(n-1)}(x) + \frac{h^n}{n!} y^{(n)}(x + \theta h),$$

$$0 \leqslant \theta \leqslant 1$$

we have

$$y(1 \cdot 0) = 1 + 1 + \frac{1}{2!} + \frac{1}{3!} + \frac{1}{4!} + \cdots \approx 2 \cdot 7$$

$$y(2 \cdot 0) = 1 + 2 + \frac{2^2}{2!} + \frac{2^3}{3!} + \frac{2^4}{4!} + \cdots \approx 7 \cdot 0$$

$$y(0 \cdot 5) = 1 + \frac{1}{2} + \frac{1}{2!}\left(\frac{1}{2}\right)^2 + \frac{1}{3}\left(\frac{1}{2}\right)^3 + \frac{1}{4!}\left(\frac{1}{2}\right)^4 + \cdots \approx 1 \cdot 65$$

The error term can be calculated and in this case it is easy to see the upper bound for the error since e^x is an increasing function and the maximum value therefore occurs for $\theta = 1$. Consider the case $y(2 \cdot 0)$ as calculated above. The error term has a maximum value of

$$\frac{h^n}{n!} e^h = \frac{2^5}{5!} e^2 \approx 2 \cdot 0$$

and many more terms would be needed to give reasonable accuracy. As suggested in the text a way of reducing the number of terms required would be to find the value of $x = 2$ by using Taylor series in stages using for example the points $0 \cdot 0, 0 \cdot 5, 1 \cdot 0, 1 \cdot 5$ in succession as the base points for the expansion.

3 Solve the following equation using Euler's method over the range $0 \leqslant x \leqslant 2 \cdot 0$ using a step length $h = 0 \cdot 5$.

$$\frac{dy}{dx} = y$$
$$y(0) = 1$$

The true solution is $y = e^x$.

$$y_{n+1} = y_n + hy_n = (1 + h)y_n$$
$$y_1 = (1 + h) \times 1 = 1 \cdot 5$$
$$y_2 = (1 + 0 \cdot 5) \times 1 \cdot 5 = 2 \cdot 25$$
$$y_3 = 1 \cdot 5 \times 2 \cdot 25 = 3 \cdot 375$$
$$y_4 = 1 \cdot 5 \times 3 \cdot 375 = 5 \cdot 062$$

It will be noted that each value depends on the last value only so that no special starting method is required.

4 Solve the above equation using the Euler–trapezoidal method with one correction only. The process is shown diagrammatically in Fig. 5.2.

$$y^p_{n+1} = y_n + hf(x_n, y_n)$$

$$y^c_{n+1} = y_n + \frac{h}{2}[f(x_n, y_n) + f(x_{n+1}, y^p_{+1})]$$

$$\begin{array}{ll}
y^p_1 = (1 + 0{\cdot}5)1 = 1{\cdot}5 & y^c_1 = 1 + 0{\cdot}25[1 + 1{\cdot}5] = 1{\cdot}625 \\
y^p_2 = (1{\cdot}5)1{\cdot}625 \approx 2{\cdot}438 & y^c_2 = 1{\cdot}625 + 0{\cdot}25[1{\cdot}625 + 2{\cdot}438] = 2{\cdot}641 \\
y^p_3 = (1{\cdot}5)2{\cdot}641 \approx 3{\cdot}962 & y^c_3 = 2{\cdot}641 + 0{\cdot}25[2{\cdot}641 + 3{\cdot}962] = 4{\cdot}292 \\
y^p_4 = (1{\cdot}5)4{\cdot}292 \approx 6{\cdot}438 & y^c_4 = 4{\cdot}292 + 0{\cdot}25[4{\cdot}292 + 6{\cdot}438] = 6{\cdot}974
\end{array}$$

Note, that in the example using Euler's method the error in y_4 is approximately $31{\cdot}5\%$ whereas in the second case the error is approximately $5{\cdot}6\%$.

5 Investigate the stability and accuracy of the formula based on Simpson's rule.

$$y_{n+2} = y_n + \frac{h}{3}[f(x_{n+2}, y_{n+2}) + 4f(x_{n+1}, y_{n+1}) + f(x_n, y_n)]$$

Consider strong stability. The auxiliary equation is $z^2 - 1 = 0$ which has roots $z = \pm 1$ and therefore the formula is not strongly unstable. Weak stability can only be investigated for a particular equation. Consider the equation

$$y' = -\lambda y, \qquad \lambda > 0$$
$$y(0) = 1$$

which has the solution $y = e^{-\lambda x}$. The auxiliary equation now becomes

$$z^2 - 1 - \frac{h}{3}[-\lambda z^2 - 4\lambda z - \lambda] = 0$$

$$z^2\left[1 + \frac{h\lambda}{3}\right] + z\left[\frac{4h\lambda}{3}\right] - 1 + \frac{h\lambda}{3} = 0$$

and the roots of this equation are

$$z = \frac{-\frac{4h\lambda}{3} \pm \sqrt{\left[\frac{16h^2\lambda^2}{9} - 4\left(-1 + \frac{h^2\lambda^2}{9}\right)\right]}}{2\left(1 + \frac{h\lambda}{3}\right)}$$

$$z_1 = \left[1 - \frac{2h\lambda}{3} + \frac{h^2\lambda^2}{6} - \frac{h^4\lambda^4}{72}\cdots\right]\left[1 - \frac{h\lambda}{3} + \frac{h^2\lambda^2}{9} - \frac{h^3\lambda^3}{27} + \frac{h^4\lambda^4}{81}\cdots\right]$$

$$= 1 - h\lambda + \frac{h^2\lambda^2}{2} - \frac{h^3\lambda^3}{6} + \frac{h^4\lambda^4}{24}\cdots$$

which agrees with the Taylor series expansion of $e^{-h\lambda}$ up to the term in h^4.

$$z_2 = \left[-1 - \frac{2h\lambda}{3} - \frac{h^2\lambda^2}{6} + \frac{h^4\lambda^4}{72}\cdots\right]\left[-\frac{h\lambda}{3} + \frac{h^2\lambda^2}{9} - \frac{h^3\lambda^3}{27} + \frac{h^4\lambda^4}{81}\cdots\right]$$

$$= -1 - \frac{h\lambda}{3} - \frac{h^2\lambda^2}{18} + \frac{h^3\lambda^3}{54} + \frac{5}{648}h^4\lambda^4\cdots$$

which agrees with the Taylor series expansion of the $-e^{h\lambda/3}$ up to the term in h^2. Therefore, for small values of h the first term approximates the solution $e^{-\lambda x}$ and the second term is a rapidly growing one which means that the formula exhibits weak stability. In the situation where the true solution is an increasing exponential, the spurious solution is decreasing and therefore of negligible importance. Thus, it can be seen that weak stability is only a problem for certain equations.

$$y(x_{n+2}) = y + 2hy' + \frac{4h^2}{2}y'' + \frac{8h^3}{6}3y''' + \frac{16h^4}{24}y'''' + \frac{32h^5}{120}y''''' + \cdots$$

$$f(x_{n+2}, y(x_{n+2})) = y' + 2hy'' + \frac{4h^2}{2}y''' + \frac{8h^3}{6}y'''' + \frac{16h^4}{24}y''''' + \cdots$$

$$f(x_{n+1}, y(x_{n+1})) = y' + hy'' + \frac{h^2}{2}y''' + \frac{h^3}{6}y'''' + \frac{h^4}{24}y''''' + \cdots$$

$$f(x_n, y(x_n)) = y'$$

The terms on the right-hand side without subscripts represent the true values at the point x_n, i.e., $y \equiv y(x_n)$. The error term is given by

$$y(x_{n+2}) - y(x_n) - \frac{h}{3}[f(x_{n+2}, y(x_{n+2})) + 4f(x_{n+1}, y(x_{n+1})) + f(x_n, y(x_n))] = E_{n+2}$$

If the various terms are added together all the terms up to h^4 have zero coefficients and E_{n+2} has a leading term $-h^5/90y''''' = -h^5/90 f''''(x_n, y(x_n))$.

6 Use the standard fourth-order Runge–Kutta formula to solve the following equation over the range $0 \leqslant x \leqslant 0.2$ with a step length $h = 0.1$.

$$y' = -2y$$
$$y(0) = 1$$

The Runge–Kutta formulae are

$$k_1 = hf(x_n, y_n)$$
$$k_2 = hf(x_n + h/2, y_n + k_1/2)$$
$$k_3 = hf(x_n + h/2, y_n + k_2/2)$$
$$k_4 = hf(x_n + h, y_n + k_3)$$

$$y_{n+1} = y_n + \tfrac{1}{6}[k_1 + 2k_2 + 2k_3 + k_4]$$

$$k_1 = 0.1(-2.0 \times 1.0) = -0.2 \qquad k_2 = 0.1(-2.0 \times 0.9) = -0.18$$
$$k_3 = 0.1(-2.0 \times 0.91) = -0.182 \qquad k_4 = 0.1(-2.0 \times 0.818) = -0.1636$$

Therefore

$$y_{0.1} = 1 + \tfrac{1}{6}[-0.2 - 0.36 - 0.364 - 0.1636] \approx 0.8187$$

$$k_1 = 0.1(-2.0 \times 0.8187) = -0.1627 \qquad k_2 = 0.1(-2.0 \times 0.7368) = -0.1474$$
$$k_3 = 0.1(-2.0 \times 0.7450) = -0.1490 \qquad k_4 = 0.1(-2.0 \times 0.6697) = -0.1339$$

$$y = 0.8187 + \tfrac{1}{6}[-0.1627 - 0.2948 - 0.2980 - 0.1339] = 0.6705$$

The true solution is $y = e^{-2x}$ which can be used to check the error. The error at $x = 0.2$ is approximately 3 in the fourth decimal place. However, if we use a value for h of 1.0 which satisfies the partial stability condition $h|\lambda| < 2.7$ the results are hopelessly inaccurate.

$$k_1 = -2.0 \times 1.0 = -2.0 \qquad k_2 = -2.0 \times 0.0 = 0.0$$
$$k_3 = -2.0 \times 1.0 = -2.0 \qquad k_4 = -2.0 \times 1.0 = +2.0$$
$$y_1 = 1 + \tfrac{1}{6}[-2.0 - 4.0 + 2.0]$$
$$= 0.3333$$

The true result is $y = 0.1353$ which gives an error of 146%. Thus, the accuracy limitation on the size of h is far more demanding than the partial stability condition.

Problems

1 Investigate the stability and accuracy of the finite-difference equation

$$y_{n+3} = y_{n+2} + \frac{h}{12}[23f_{n+2} - 16f_{n+1} + 5f_n]$$

2 Consider which method or methods you would use in the following circumstances.

(a) The derivative function is complicated and requires special programs for its evaluation.

(b) The derivative function varies in magnitude considerably at different points in the range.

(c) Two simultaneous equations are to be solved.

$$y' = -10y + 6z, \qquad y(0) = 4$$
$$z' = 13 \cdot 5y - 10z, \qquad z(0) = 0$$

3 Solve Worked Example 4 by iterating to convergence to four decimal places and compare the accuracy of this answer with the one obtained with one iteration only.

4 Write a program to solve the equation

$$y' = \sqrt{(1 - y^2)}$$
$$y(0) = -1$$

using the Adams predictor–corrector method given in Eqns (5.25) and (5.26) with two corrections over the range $0 \times 0 \cdot 9$ with $h = 0 \cdot 1$.

5 Solve the above equation with one application of the corrector followed by a step to remove the error term as described in Section 5.3.3. Compare the results of this calculation and the one above.

6 Consider the formula

$$y_{n+2} = \alpha_1 y_{n+1} + \alpha_0 y_n + h\beta_1 f_{n+1}$$

and find the values of α_1, α_0, and β_1 to give the maximum order of accuracy. Give the leading error term and comment on stability properties of the formula.

6

Finite differences

6.1 Behaviour of finite differences

Thirty years ago the work involved in using numerical methods was excessive since calculations were performed either by slide rule or with the aid of mechanical desk calculators. In this environment there was an emphasis on methods which were simple arithmetically, and suitable for pencil and paper calculations. The methods based on differences were well suited to this situation and were extensively used.

With the advent of electronic digital computers with very high speeds of operation, but low storage capacity, the storage of large tables of finite differences was less feasible and methods formulated in other ways were developed. The present situation has changed yet again since storage limitations are not a major problem in many scientific computing establishments. This means that finite-difference formulations are again practical and may regain a more significant place in numerical methods.

Finite-difference methods are applied to functions for which values are available at equidistant points. There are a series of points $x_n = x_0 + n \cdot h$ $(n = 0, 1, \ldots, N)$ with corresponding function values f_n $(n = 0, 1, \ldots, N)$.

By subtracting successive function values a column of first-order differences can be drawn up

$$\Delta f_r = f_{r+1} - f_r \tag{6.1}$$

In a similar manner a column of second-order differences can be calculated from the formula

$$\Delta^2 f_r = \Delta f_{r+1} - \Delta f_r \tag{6.2}$$

Higher-order differences can be found in a similar manner. The finite-difference table gives a useful method for approximating a function when the function can be represented approximately by a polynomial. In such a case the columns of differences become negligible after a certain point and only the first columns of differences need be included in the calculations.

For example, Table 6.1 showing the differences for a cubic has the fourth column of differences and all subsequent columns equal to zero.

In the case of a polynomial of degree n, the $n + 1$th column of differences and all subsequent columns are zero. The reverse is also true; namely, that if a column of differences is found to be zero then the function values lie on a polynomial of degree equal to the highest order of differences which are nonzero. In the case of a practical example, however, the columns of differences may never become exactly zero owing to the presence of round-off error. Most frequently, the function concerned is not a polynomial but is to be approximated by some suitable polynomial. In such a case the difference table is built up and the various columns inspected. If column k of the difference table is sufficiently small to be considered negligible then the difference table is truncated at this point. This is equivalent to approximation by a polynomial of degree $k-1$.

Table 6.1
Difference table for a cubic

x_n	f_n	Δ	Δ^2	Δ^3	Δ^4
1	1				
		7			
2	8		12		
		19		6	
3	27		18		0
		37		6	
4	64		24		0
		61		6	
5	125		30		0
		91		6	
6	216		36		
		127			
7	343				

However, a further example will show that this behaviour does not take place for all functions. Consider the function $f(x) = 2^x$ shown in Table 6.2.

Table 6.2
Difference table for 2^x

x_n	f_n	Δ	Δ^2	Δ^3
1	2			
		2		
2	4		2	$\cdots$
		4		
3	8		4	$\cdots$
		8		
4	16		8	$\cdots$
		16		
5	32		16	$\cdots$
		32		
6	64			

In the above case all columns of differences are significant and none of them can be ignored. Table 6.3 illustrating differences of $\sqrt{x}$ shows another type of behaviour where the columns of differences become smaller but a systematic pattern remains.

Table 6.3

Difference table for $x^{\frac{1}{2}}$

x_n	f_n	Δ	Δ^2	Δ^3	Δ^4
1	1·000				
		0·414			
2	1·414		−0·096		
		0·318		0·046	
3	1·732		−0·050		−0·028
		0·268		0·018	
4	2·000		−0·032		−0·009
		0·236		0·009	
5	2·236		−0·023		−0·002
		0·213		0·007	
6	2·449		−0·016		
		0·197			
7	2·646				

In the subsequent discussion in this chapter it will be assumed that the functions considered can be adequately approximated by a polynomial.

6.2   Errors in finite-difference tables

Most function values cannot be represented exactly in a finite number of digits so that the table of function values contains some error due to the round-off error. Errors may also be present owing to clerical errors such as transposition of digits or repetition of the wrong digit, e.g., 977 instead of 997. It is instructive to see how this error builds up as the various columns of differences are calculated. This can be shown by considering a difference table in which every entry is zero and introducing an error e in one function value. Table 6.4 shows the growth of error.

Table 6.4

Error build-up in difference table

f_n	Δ	Δ^2	Δ^3	Δ^4
0				
	0			
0		0		
	0		0	
0		0		+e
	0		+e	
0		+e		−4e
	+e		−3e	
e		−2e		6e
	−e		+3e	
0		+e		−4e
	0		−e	
0		0		+e
	0		0	
0		0		
	0			
0				

The coefficients in each column are the binomial coefficients $_nC_r = n!/(n-r)!\,r!$ which grow in size as n increases.

One of the problems with finite differences is the way the errors are magnified by the differencing process, but a knowledge of the pattern generated by the errors may enable errors to be identified and eliminated. Consider Table 6.5.

Table 6.5

A difference table with an error

x_n	f_n	Δ	Δ^2	Δ^3	Δ^4	Δ^5
$x_0 = 0$	100					
		1				
0·5	101		14			
		15		36		
1·0	116		50		24	
		65		60		0
1·5	181		110		24	
		175		84		−27
2·0	356		194		−3	
		369		81		135
2·5	725		275		132	
		644		213		−270
3·0	1369		488		−138	
		1132		75		+270
3·5	2501		563		132	
		1695		207		−135
4·0	4196		770		−3	
		2465		204		
4·5	6661		974			
		3439				
5·0	10100					

The ratio of the terms in the last column is $1:5:10:10:5$ which implies that the error occurred five columns back, since these are the binomial coefficients $_5C_r$. The size of the error is -27 and by following the diagonal sloping down to the first column the position of the error can be determined exactly. There is an error of -27 in the entry 1369. This can easily be checked by using the formula $f(x) = 100 + 16x^4$ which gives the results of Table 6.5. Two further examples of error propagation in a finite-difference table are given in Examples 6.1 and 6.2.

Thus, where it is known that a function approximates to a polynomial it is possible to trace errors in this way. The problem is more difficult when rounded numbers are used since the errors will not then be exact multiples of the binomial coefficients.

6.3 Finite-difference operators

We have already used the forward-difference operator defined by

$$\Delta f_r = f_{r+1} - f_r \tag{6.3}$$

The operator is so called because the operation is defined using the next highest point of the sequence and the present one to define the difference. In a similar manner the backward-difference operator is defined by

$$\nabla f_r = f_r - f_{r-1} \tag{6.4}$$

Other operators which are used require the use of subscripts at the midpoints. The central-difference operator is defined by

$$\delta f_{r+1/2} = f_{r+1} - f_r \tag{6.5}$$

The averaging operator is defined by

$$\mu f_{r+1/2} = \tfrac{1}{2}[f_{r+1} + f_r] \tag{6.6}$$

In order to relate these operators we require a further operator, the shift operator, defined by

$$E f_r = f_{r+1} \tag{6.7}$$

When using operators it is often convenient to consider the algebra of the operators divorced from the function values on which they operate. In such circumstances it is necessary to analyse very carefully the conditions under which this algebra is applicable. This analysis is presented in the standard texts on finite differences such as Hildebrand (1956) or Redish (1961).

Some results will be quoted assuming that the appropriate conditions, which are required to ensure their validity, are satisfied.

1 $\Delta f_r = f_{r+1} - f_r$

Hence,

$$\Delta \equiv E - 1$$
$$E \equiv 1 + \Delta \tag{6.8}$$

2 $\nabla f_r = f_r - f_{r-1}$

Hence,

$$\nabla \equiv 1 - E^{-1}$$
$$E \equiv (1 - \nabla)^{-1} \tag{6.9}$$

3 $\delta f_{r+1/2} = f_{r+1} - f_r$

Hence,

$$\delta \equiv E^{1/2} - E^{-1/2} \tag{6.10}$$

4 From Eqns (6.8) and (6.9)

$$1 + \Delta \equiv \frac{1}{1 - \nabla}$$

Therefore,

$$\Delta \equiv \frac{1}{1 - \nabla} - 1 \equiv \frac{\nabla}{1 - \nabla} \tag{6.11}$$

$$\nabla \equiv 1 - \frac{1}{1 + \Delta} \equiv \frac{\Delta}{1 + \Delta} \tag{6.12}$$

There are many more relationships of this type which the reader can derive for himself. Some of the above relationships will be used to derive some of the standard finite-difference relationships in later sections.

It is also useful to have formulae relating higher-order differences to function values. For example,

$$\Delta^2 f_r = \Delta f_{r+1} - \Delta f_r$$
$$= f_{r+2} - f_{r+1} - (f_{r+1} - f_r)$$
$$= f_{r+2} - 2f_{r+1} + f_r$$
$$= (E - 1)^2 f_r \qquad (6.13)$$

and generally

$$\Delta^r = (E - 1)^r \qquad (6.14)$$

If we refer to Table 6.5 we can see which elements correspond to the various differences. A sequence of forward differences with the same subscript occur down a forward-sloping diagonal. For example,

$$f_2 = 116, \qquad \Delta f_2 = 65, \qquad \Delta^2 f_2 = 110, \qquad \Delta^3 f_2 = 84, \quad \text{etc.}$$

The backward differences with constant subscript are found on a backward-sloping diagonal.

$$f_4 = 356, \qquad \nabla f_4 = 175, \qquad \nabla^2 f_4 = 110, \qquad \nabla^3 f_4 = 60$$

Note, that the element 110 is both $\Delta^2 f_2$ and $\nabla^2 f_4$. The central differences with constant subscript are found on a horizontal line

$$f_3 = 181, \qquad \delta^2 f_3 = 110, \qquad \delta^4 f_3 = 24$$

Examples 6.3 and 6.4 show further formulae relating the various operators and the function values.

6.4 Interpolation

By an extension of the above notation (which is justified in mathematical texts) it is possible to derive formulae which will estimate the value of the function at a non-tabular point. In the case of a table such as Table 6.1 or the corrected version of Table 6.5 the differences eventually become zero and therefore, formulae can be used which are exact, apart from rounding errors.

However, with a table such as Table 6.3, the finite differences of high order will be small but not zero. In such a case a finite number of difference columns will be used and this introduces an error due to the truncation of the sequence of finite differences. If, for example, Table 6.3 is truncated after the third column of differences then the approximation will be equivalent to approximation by a cubic over a particular set of four points.

The extension of the formula is derived as follows. If p and k are integers

$$f_{p+k} = E f_{p+k-1}$$
$$= E^2 f_{p+k-2}$$
$$= E^p f_k \qquad (6.15)$$

Now, let us assume the formula still holds when f_k is a tabular point where $x_k = x_0 + k.h$ but p is a fractional value so that f_{p+k} is a non-tabular point. Using Eqn (6.8), $E \equiv 1 + \Delta$, we have

$$f_{p+k} = E^p f_k$$
$$\equiv (1 + \Delta)^p f_k$$
$$= \left(1 + p\Delta + \frac{p.(p-1)}{2!}\Delta^2 \cdots\right) f_k$$
$$= f_k + p\,\Delta f_k + \left(\frac{p^2 - p}{2}\right)\Delta^2 f_k \cdots \qquad (6.16)$$

The formula will terminate if the function values represent a polynomial.

This formula is known as Newton's forward-difference formula and uses values down a forward diagonal of a difference table. It is therefore particularly useful at the start of a table when there are not sufficient central-difference values on a horizontal line (or backward differences along a backward-sloping line) to use in a formula.

A similar formula using backward differences is easily found by the same reasoning.

$$f_{p+k} = E^p f_k$$
$$= (1 - \nabla)^{-p} f_k$$
$$= (1 + p\nabla + \frac{p(p+1)}{2!}\nabla^2 \cdots) f_k$$
$$= 1 + p\nabla f_k + \frac{(p^2 + p)}{2}\nabla^2 f_k \cdots \qquad (6.17)$$

This formula is particularly suitable for interpolation at the end of a table where only backward differences are available.

In texts which include mathematical analysis of finite-difference methods (see Hildebrand (1956) or Hartree (1958)) it is shown that these forward- or backward-difference formulae are not as accurate as formulae based on central differences. There are central-difference formulae known by the names of Bessel, Stirling, and Everett and one of these formulae would normally be used where the appropriate differences were available.

The use of a formula such as Newton's forward-difference formula is best shown by means of an example. If we wish to find the value of $f(0.7)$ using Newton's forward-difference formula based on the point $x = 0.5$ we have using the data in Table 6.5

$$x_{p+k} = x_k + p \times h$$
$$0.7 = 0.5 + p \times 0.5$$

Therefore,

$$p = 0.4 \qquad (6.18)$$

Hence, successive approximations are

$$101 + 0.4 \times 15 = 107$$

$$101 + 0.4 \times 15 \frac{+ (0.4) \times (-0.6) \times 50}{2!} = 101$$

(6.19)

$$101 + 6.0 - 6.0 \frac{+ (0.4)(-0.6)(-1.6) \times 60}{6} = 104.84$$

$$101 + 6.0 - 6.0 + 3.84 \frac{+ (0.4)(-0.6)(-1.6)(-2.6) \times 24}{4 \times 6} = 103.8416$$

The latter is the exact value since the function is a polynomial of degree 4, i.e., $100 + 16x^4$ and all differences up to the fourth have been included.

6.5 Formulae for differentiation and integration

Before giving examples of the use of finite-difference formulae for differentiation and integration it is perhaps worthwhile looking at the problem graphically to gain appreciation of the difficulties involved. Consider the two examples in Fig. 6.1(a) and (b).

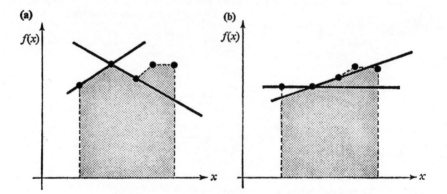

Fig. 6.1 Error in integral and derivative.

The difference between the two figures is that the second point has a slightly different function value. The difference has been exaggerated so that it could easily be seen on the diagram but values will normally be subject to some round-off or experimental error. The effect of this error on the approximate derivative through two adjacent points is dramatic, as shown in the diagram. However, if the integral is approximated by the enclosed area, we see that the error in the integral is relatively small.

These crude insights can be supported by mathematical analysis of finite-difference methods. This shows that differentiation of finite-difference formulae can lead to large errors, whereas the integration process is basically well-conditioned. The formulae for differentiation will, therefore, not be pursued. Where possible, problems should be formulated so that differentiation of finite-difference formulae, or approximation functions, is not required.

Useful formulae for integration can be derived, for example, by integration of Newton's forward-difference formula. Since, $x_{p+k} = x_k + p.h$ where p is a variable and x_k a tabular point, we have

$$\frac{dx}{dp} = h \tag{6.20}$$

$$\int_{x_k}^{x_{k+1}} f(x)\,dx = h\int_0^1 f_{p+k}\,dp$$

$$= h\int_0^1\left[1 + p\Delta + \frac{p(p-1)}{2}\Delta^2\cdots\right]f_k\,dp$$

$$= h\left[pf_k + \frac{p^2}{2}\Delta f_k + \left(\frac{p^3}{6} - \frac{p^2}{4}\right)\Delta^2 f_k\cdots\right]_0^1$$

$$= h[f_k + \tfrac{1}{2}\Delta f_k - \tfrac{1}{12}\Delta^2 f_k\cdots] \tag{6.21}$$

The first two terms of this series give

$$I \approx \frac{h}{2}[f_k + f_{k+1}] \tag{6.22}$$

which is the trapezoidal formula for approximate integration. By integrating from x_k to x_{k+2} and taking the first three terms the formula

$$I \approx h[2f_k + 2\Delta f_k + \tfrac{1}{3}\Delta^2 f_k]$$

$$= \frac{h}{3}[f_k + 4f_{k+1} + f_{k+2}] \tag{6.23}$$

is obtained which is the familiar Simpson's rule for integration.

There are many different possible formulae which can be derived by these means by using the various difference formulae and integrating over a particular number of steps. Exhaustive examples of these formulae are available in texts on finite differences but a more general method of generating formulae will be used in Chapter 8 on numerical integration.

6.6 Solution of linear difference equations

In the study of differential equations, the solution of a linear difference equation determines the stability of a method. We therefore consider the solutions of an equation of the form

$$\alpha_k y_{n+k} + \cdots + \alpha_1 y_{n+1} + \alpha_0 y_n = 0 \tag{6.24}$$

This is a linear difference equation of degree k which has k linearly independent solutions. The complete solution is a linear combination of these solutions and therefore involves k arbitrary constants. These constants are found by k extra conditions.

$$y_r = s_r, \qquad r = 0, 1, \ldots, k-1 \tag{6.25}$$

Solutions of Eqn (6.24) can be found by substituting a trial solution $y_n = A(z)^n$ and finding values of z such that the equation is satisfied. Substituting the trial solution gives

$$Az^n[\alpha_k z^k + \alpha_{k-1}z^{k-1} + \cdots + \alpha_0] = 0 \tag{6.26}$$

This polynomial is known as the auxiliary polynomial and any root z_r of this polynomial leads to a solution $y_n = A(z_r)^n$ of the difference equation. A complication arises if there are multiple roots of the difference equation since not all the roots can then have this simple form. In the case of a double root z_r this leads to two roots, z_r and $n \cdot z_r$. For a triple root the three independent solutions are z_r, nz_r, and $n^2 z_r$, etc.

Thus, when there are n distinct roots the complete solution is

$$y_n = A_1(z_1)^n + A_2(z_2)^n + \cdots + A_k(z_k)^n \tag{6.27}$$

If z_1 were a double root this would become

$$y_n = (A_1 + A_2 n)(z_1)^n + A_3(z_2)^n + \cdots + A_k(z_k)^n \tag{6.28}$$

The k initial conditions of Eqn (6.25) are then sufficient to fix the constants A_r $(r = 1, \ldots, k)$ by solving a set of linear simultaneous equations. In the case of distinct roots these are

$$
\begin{aligned}
A_1 + A_2 + \cdots + A_k &= s_0 \\
A_1 z_1 + A_2 z_2 + \cdots + A_k z_k &= s_1 \\
&\cdots\cdots\cdots\cdots\cdots\cdots\cdots \\
A_1 z_1^{k-1} + A_2 z_2^{k-1} + \cdots + A_k z_k^{k-1} &= s_{k-1}
\end{aligned}
\tag{6.29}
$$

A worked example is given in Example 6.5.

Bibliographic notes

There are many well-established texts containing extensive material on finite-difference methods. The books by Hartree (1957), Hildebrand (1956), Scarborough (1958), and Buckingham (1957) are examples of these. There are also texts containing large numbers of worked numerical examples such as those of Redish (1961), Butler and Kerr (1962), and Bull (1966). The book by Lanczos considers the problem of differentiation of a finite-difference formula and gives a possible method of solution. The solution of difference equations is discussed in the book by Goldberg (1958).

Worked examples

1 The following table shows another way that errors can build up in a difference table if the errors are alternately positive and negative. This pattern which gives the maximum build-up of errors shows the errors superimposed on a table with all entries zero.

$+\varepsilon$					
	-2ε				
$-\varepsilon$		$+4\varepsilon$			
	$+2\varepsilon$		-8ε		
$+\varepsilon$		-4ε		$+16\varepsilon$	
	-2ε		$+8\varepsilon$		-32ε
$-\varepsilon$		$+4\varepsilon$		-16ε	
	$+2\varepsilon$		-8ε		
$+\varepsilon$		-4ε			
	-2ε				
$-\varepsilon$					

2 A table in which two errors have been made is more difficult to analyse since the binomial coefficients overlap. The following pattern shows a possible example. It

should be realized that in a real example the function values will be significant in the first few columns and the errors will have to be traced from the right-hand columns where the maximum interaction has taken place.

f	Δ	Δ^2	Δ^3	Δ^4
0				
	0			
0		0		
	0		ε_1	
0		ε_1		$-4\varepsilon_1$
	ε_1		$-3\varepsilon_1$	
ε_1		$-2\varepsilon_1$		$6\varepsilon_1$
	$-\varepsilon_1$		$3\varepsilon_1$	
0		ε_1		$\varepsilon_2-4\varepsilon_1$
	0		$\varepsilon_2-\varepsilon_1$	
0		ε_2		$+4\varepsilon_2+\varepsilon_1$
	ε_2		$-3\varepsilon_2$	
ε_2		$-2\varepsilon_2$		$6\varepsilon_2$
	$-\varepsilon_2$		$3\varepsilon_2$	
0		ε_2		
	0			
0				

It may be possible to see the error pattern in the third-order differences but the confusion in the fourth difference column would probably be too great to give an opportunity to detect the errors.

3 The relationship between a formula expressed in terms of differences and the same formula expressed as a sum of function values can easily be established using the various formulae relating the different operators. For example,

$$\nabla = 1 - E^{-1}, \quad \nabla^2 = E^{-2} - 2E^{-1} + 1, \quad \text{etc.}$$

The Adams–Bashforth formula which is used for the integration of ordinary differential equations is

$$\nabla y_{n+1} = h[1 + \tfrac{1}{2}\nabla + \tfrac{5}{12}\nabla^2]y_n$$

up to second-order differences. This can be expressed

$$y_{n+1} - y_n = hy_n' + \frac{h}{2}[y_n' - y_{n-1}'] + \frac{5h}{12}[y_n' - 2y_{n-1}' + y_{n-2}']$$

$$= \frac{23h}{12}y_n' - \frac{16h}{12}y_{n-1}' + \frac{5h}{12}y_{n-2}'$$

4 Find the relation between the operators E and δ.

$$\delta f_r = f_{r+1/2} - f_{r-1/2}$$

therefore,

$$\delta = E^{1/2} - E^{-1/2}$$
$$\delta^2 = E - 2 + E^{-1}$$
$$E - 2 - \delta^2 + E^{-1} = 0$$
$$E = 2 + \frac{\delta^2 + \sqrt{[(2+\delta^2)^2 - 4]}}{2}$$

5 Solve the linear difference equation

$$y_{r+4} - y_{r+3} - 3y_{r+2} + y_{r+1} + 2y_r = 0$$

with the initial conditions $y_0 = 1$, $y_1 = -1$, $y_2 = 2$, and $y_3 = 7$. We form the auxiliary equation

$$z^4 - z^3 - 3z^2 + z + 2 = 0$$

By inspection there is one root $z = 1$. Therefore, we look for roots of the cubic found by division by $z - 1$.

$$z^3 - 3z - 2 = 0$$

This has solutions $z = -1$ which is a double root and $z = 2$. The general solution is, therefore,

$$y_n = A(+1)^n + B(-1)^n + Cn(-1)^n + D(2)^n$$

Using the initial conditions gives

$$A + B + D = 1$$
$$A - B - C + 2D = -1$$
$$A + B + 2C + 4D = 2$$
$$A - B - 3C + 8D = 7$$

Adding the first and second equations and subtracting the sum of the third and fourth equations gives

$$-9D = -9$$

therefore,

$$D = 1, \quad A + B = 0$$

From the second and third equations $3A - B = -8$ and, therefore,

$$A = -2, \quad B = 2 \quad \text{and} \quad C = -4 + 4 + 1 = 1$$

Problems

1 Find the error in this table and correct it.

x	0·0	0·1	0·2	0·3	0·4	0·5
$f(x)$	4·000	4·641	5·368	6·187	7·104	8·125

x	0·6	0·7	0·8	0·9	1·0
$f(x)$	9·265	10·503	11·872	13·369	15·000

2 Find the relationship between the operators μ and δ

3 Express the following closed Adams–Bashforth formula up to second differences in terms of function values

$$\nabla y_{n+1} = h[1 - \tfrac{1}{2}\nabla - \tfrac{1}{12}\nabla^2]y'_{n+1}$$

4 Solve the linear difference equation

$$2y_{r+3} + y_{r+2} - 2y_{r+1} - y_r = 0$$

with the initial conditions $y_0 = -1$, $y_1 = 2$, and $y_2 = \frac{1}{2}$.

5 Integrate Newton's backward-difference formula up to third-order differences and obtain various formulae expressed in terms of function values by truncating the formula after the first difference, etc.

7

Curve fitting

7.1 Interpolation and approximation

There are two types of problem which will be considered under this heading; firstly, the problem of interpolation which involves finding intermediate values when values are given at a finite set of points and, secondly, the problem of approximation to a function over a complete range of values by a simple function which is more suitable for computation. Clearly, the object of our approximation should be to make the error as small as possible, and different methods arise corresponding to different ways of defining the error.

In the case of interpolation by finite-difference methods, the approximating function is a polynomial which has values equal to the given values at a finite set of points. Let f_i be the given values at the points x_i $(i = 0, 1, \ldots n)$ and let $\phi_n(x)$ be the approximating function. If we define the error by

$$E_n = \sum_{i=0}^{n} |\phi_n(x_i) - f_i| \tag{7.1}$$

then fitting the function exactly at the $n+1$ points will reduce the error E_n to zero. The error defined in Eqn (7.1) has certainly been minimized, but the question remains whether the values at points where $x \neq x_i$ are good approximations.

In the approximation problem we are concerned with the error at all points in the range. The problems involved in using an error definition which depends only on the values at a finite set of points are shown very clearly by an example discovered by Runge in 1901. This example is discussed in Lanczos (1957). A set of equidistant points, x_i $(i = 0, 1, \ldots, n)$ are chosen to approximate the function $1/(1 + 25x^2)$ over the range $[-1, +1]$. It is found that for any point $x \neq x_i$, where $|x| > 0.726$, the error $|\phi_n(x) - f(x)|$ of the approximation increases without bound as n increases, and this is true even though $\phi_n(x_i) = f(x_i)$ $(i = 0, 1, \ldots, n)$ which means that $E_n = 0$.

When considering the error over a whole range, a more satisfactory objective is to make the maximum error as small as possible. This is the minimax type of approximation where the error is defined by

$$E_{\max} = \max_{a \leqslant x \leqslant b} |\phi(x) - f(x)| \tag{7.2}$$

and the function $\phi(x)$ is chosen so that $E_{\max}$ is minimized. It is in this context that the Chebyshev polynomials have found wide application.

The third case which is of interest is where the number of points at which values are given is considerably greater than the degree of approximation which is desirable. For example, it may be desirable to use a low-order polynomial, say a cubic, as an approximation over a range in which perhaps twelve function values are known. Four points would be sufficient to determine a cubic uniquely and errors would, therefore, arise at the remaining points.

In this situation, rather than make the error zero at any particular point, we require the overall error to be as low as possible. An appropriate choice of error definition is given by

$$S_m = \sum_{i=0}^{m} [\phi_n(x_i) - f_i]^2, \qquad m \geqslant n \tag{7.3}$$

The least-squares fit is obtained by finding a function $\phi_n(x)$ which minimizes the quantity S_m. The subscript n implies that the function $\phi_n(x)$ is dependent on a number of parameters which can be given appropriate values to obtain the least-squares property. In the case of a polynomial these parameters are the coefficients $a_0, a_1, \ldots, a_n$, and the function $\phi_n(x)$ would have $n+1$ variable parameters.

Polynomials are widely used for approximation so it is worthwhile considering whether they are the most appropriate functions for this purpose. One big advantage is that by using the arithmetic operations available on a digital computer, it is possible to evaluate directly a polynomial or the quotient of two polynomials. However, the calculation of other functions, such as exponential or trigonometric functions is by means of approximation methods. Also, it is easy to evaluate both integrals and derivatives of polynomials by direct calculation.

It is also important to know how close an approximation can be obtained using polynomials. Fortunately, there is a theorem due to a Weierstrass which shows that, for any continuous function on a finite interval, the minimax error can be made as small as we please by choosing a polynomial of sufficiently high degree.

Another type of approximation which is also significant is approximation by Fourier series. In this case it can be shown that arbitrarily close approximation can be obtained for a much wider class of functions, such as those satisfying the Dirichlet conditions. (See Lanczos (1957).)

These different forms of approximation are now described. A discussion on the properties of orthogonal polynomials is also included in view of their importance in approximation theory.

7.2 Interpolation

7.2.1 Lagrange interpolation

It was shown in the previous chapter that finite-difference tables could be used for interpolation, but this was restricted to the case of function values at equidistant intervals. The Lagrange form of the interpolation polynomial gives a polynomial which fits a given function exactly at a number of points with arbitrary position, i.e.,

$$P_n(x_i) = f_i, \qquad i = 0, 1, \ldots, n \tag{7.4}$$

or in expanded form

$$a_0 + a_1x_0 + \cdots + a_nx_0^n = f_0$$
$$a_0 + a_1x_1 + \cdots + a_nx_1^n = f_1$$
$$\cdots\cdots\cdots\cdots\cdots\cdots\cdots\cdots\cdots$$ (7.5)
$$a_0 + a_1x_n + \cdots + a_nx_n^n = f_n$$

If all the x_i are distinct, then the matrix associated with these equations is known to be non-singular and therefore the equations can theoretically be solved to find the coefficients a_i. However, the equations can be ill-conditioned and it is, therefore, preferable to find the coefficients by another method.

We can see from the equations that the solutions a_i are linear combinations of the f_i. If this type of expression for the a_i is inserted in the polynomial, and the terms in each f_i are gathered together, we have

$$P_n(x) = \sum_{i=0}^{n} l_{n,\,i}(x)f_i$$ (7.6)

where the $l_{n,\,i}(x)$ are polynomials of degree n at the most.

It is easy to deduce the form of the polynomials $l_{n,\,i}(x)$ by using Eqns (7.4) which define Lagrange interpolation. At the point x_0 all the values $l_{n,\,i}(x_0)$ $(i = 1, 2, \ldots, n)$ must equal zero in order that the corresponding f_i terms are eliminated. This gives a root at $x = x_0$ for all polynomials $l_{n,\,i}(x_0)$ $(i \neq 0)$. Also, $l_{n,\,0}(x_0)$ must equal unity to give the correct coefficient of f_0. The same argument can be applied to each point, showing that each polynomial has a root at the other n points of the set.

For example,

$$l_{n,\,0}(x_0) = A_0(x - x_1)(x - x_2)\cdots(x - x_n)$$ (7.7)

and generally

$$l_{n,\,i}(x) = A_i(x - x_0)\cdots(x - x_{i-1})(x - x_{i+1})\cdots(x - x_n)$$ (7.8)

The second condition, which gives the value at the point x_i, enables the constant A_i to be found, i.e., since $l_{n,\,i}(x_i) = 1$ then

$$A_i = \frac{1}{(x_i - x_0)\cdots(x_i - x_{i-1})(x_i - x_{i+1})\cdots(x_i - x_n)}$$ (7.9)

The polynomials can thus be written concisely

$$l_{n,\,i}(x) = \prod_{r=0}^{n}{}' \frac{x - x_r}{x_i - x_r}$$ (7.10)

The prime is used to indicate that the x_i term is omitted from the product.

If we wish to interpolate at any point x, the values of the polynomials are evaluated and then multiplied by the corresponding values f_i. For example, the interpolated value of $f(5)$ can be found from the following table as follows.

	x_0	x_1	x_2	x_3
x_i	1	4	7	9
f_i	2	13	122	504

$$l_{3,\,0}(x) = \frac{(x - 4)(x - 7)(x - 9)}{(1 - 4)(1 - 7)(1 - 9)}$$ (7.11)

E

Hence,

$$l_{3,0}(5) = \frac{1.(-2).(-4)}{-3.(-6).(-8)} = -\frac{1}{18} \tag{7.12a}$$

and similarly,

$$l_{3,1}(5) = \frac{(5-1)(5-7)(5-9)}{(4-1)(4-7)(4-9)} = \frac{32}{45} \tag{7.12b}$$

$$l_{3,2}(5) = \frac{(5-1)(5-4)(5-9)}{(7-1)(7-4)(7-9)} = \frac{4}{9} \tag{7.12c}$$

$$l_{3,3}(5) = \frac{(5-1)(5-4)(5-7)}{(9-1)(9-4)(9-7)} = -0.1 \tag{7.12d}$$

Hence,

$$P_n(5) = -\tfrac{1}{18}.2 + \tfrac{32}{45}.13 + \tfrac{4}{9}.122 - 0.1.504 = 12.95 \tag{7.13}$$

The above calculations are rather complicated, and would lead to considerable computation if interpolated values were required at a large number of points. Also, it can be seen that introduction of further points into the scheme, in an effort to increase the accuracy, would change all the coefficients, therefore, any previous computation would be wasted. For these reasons the Lagrange method is not the most suitable form for interpolation on the computer, although it had wide popularity as a desk-machine method.

7.2.2   Divided differences
It is proved in most mathematical texts on numerical analysis that if we have a polynomial satisfying the conditions

$$P_n(x_i) = f_i, \qquad i = 0, 1, \ldots, n \tag{7.14}$$

then this polynomial is unique, provided the points x_i are distinct, (e.g., see Isaacson and Keller (1966)). Therefore, the problem is to rearrange the lagrangian polynomial in a form more suitable for automatic computation. In particular it should be possible to add further values to the scheme in a simple manner, without invalidating the previous computation.

Assume that we have an interpolation polynomial $P_k(x)$ of degree at most k, which fits the data at the points, x_i $(i = 0, 1, \ldots, k)$ and we wish to form the next member $P_{k+1}(x)$ by adding a further interpolation point x_{k+1}. Since it is required that the previous calculations do not need to be altered, we seek a form

$$P_{k+1}(x) = P_k(x) + p_{k+1}(x), \qquad k = 0, 1, \ldots, n-1 \tag{7.15}$$

where $p_{k+1}(x)$ has degree $k+1$ at the most.

Since $P_{k+1}(x)$ and $P_k(x)$ interpolate at the points x_i $(i = 0, 1, \ldots, k)$ we have

$$P_{k+1}(x_i) = P_k(x_i), \qquad i = 0, 1, \ldots, k \tag{7.16}$$

and hence, $p_{k+1}(x_i) = 0$ using Eqn (7.15). Since all the $k+1$ factors are revealed by this analysis the form of $p_{k+1}(x)$ is

$$p_{k+1}(x) = a_{k+1}(x - x_0)(x - x_1) \cdots (x - x_k) \tag{7.17}$$

The interpolation polynomial which results is known as Newton's divided difference interpolation polynomial and has the form

$$p_n(x) = a_0 + a_1(x - x_0) + \cdots + a_n(x - x_0)(x - x_1) \cdots (x - x_{n-1}) \quad (7.18)$$

This can be expressed in the nested form which is convenient for computation, e.g.,

$$P_3(x) = [[a_3(x - x_2) + a_2](x - x_1) + a_1](x - x_0) + a_0 \quad (7.19)$$

Fortunately, the coefficients a_k can be generated quite simply by building up a table rather similar to a finite difference table. By substituting the values x_i $(i = 0, 1, \ldots, n - 1)$ in Eqn (7.18) and using $P(x_i) = f_i$, we have

$$f_0 = a_0$$
$$f_1 = a_0 + (x_1 - x_0)a_1$$
$$f_2 = a_0 + (x_2 - x_0)a_1 + (x_2 - x_0)(x_2 - x_1)a_2$$
$$\cdots\cdots\cdots\cdots\cdots\cdots\cdots\cdots\cdots\cdots\cdots\cdots\cdots\cdots \quad (7.20)$$

The coefficients a_i are given by

$$a_0 = f_0 \quad (7.21a)$$

$$a_1 = \frac{f_1 - f_0}{x_1 - x_0} \quad (7.21b)$$

This term is called a first-order divided difference and is denoted by $f[x_1, x_0]$ or f_{01}. The second-order difference for the points x_2, x_1, x_0 is found from Eqns (7.20)

$$a_2 = \frac{(f_2 - f_0)/(x_2 - x_0) - f[x_1, x_0]}{x_2 - x_1} = \frac{f[x_2, x_0] - f[x_1, x_0]}{x_2 - x_1} \quad (7.21c)$$

and is denoted by $f[x_2, x_1, x_0]$ or f_{012}. The other coefficients are given by higher-order divided differences

$$a_n = f[x_n, x_{n-1}, \ldots, x_0] \equiv f_{01}, \ldots, {}_n \quad (7.22)$$

Various properties of divided differences and associated theorems are presented in Hildebrand (1956) and Isaacson and Keller (1966). It is sufficient here to note that alteration of the order of the terms does not change the value of a divided difference, and that the error term for interpolation using $P_n(x)$ at an arbitrary point x is

$$E(x) = (x - x_0)(x - x_1) \cdots (x - x_n)f[x_0, x_1, \ldots, x_n, x] \quad (7.23)$$

It can be shown that

$$f[x_0, x_1, \ldots, x_n, x] = \frac{f^{(n+1)}(\zeta)}{(n + 1)!} \quad (7.24)$$

where ζ is some point in the interval in which the points x_i lie.

The coefficients are calculated by building up a divided difference table as shown

$$
\begin{array}{ll}
x_0 \quad f_0 \\
\qquad \dfrac{f_1 - f_0}{x_1 - x_0} \equiv f_{01} \\
x_1 \quad f_1 \qquad\qquad\qquad \dfrac{f_{12} - f_{01}}{x_2 - x_0} \equiv f_{012} \\
\qquad \dfrac{f_2 - f_1}{x_2 - x_1} \equiv f_{12} \\
x_2 \quad f_2 \qquad\qquad\qquad \dfrac{f_{23} - f_{12}}{x_3 - x_1} \equiv f_{123} \\
\qquad \dfrac{f_3 - f_2}{x_3 - x_2} \equiv f_{23} \\
x_3 \quad f_3 \qquad\qquad\qquad \dfrac{f_{34} - f_{23}}{x_4 - x_2} \equiv f_{234} \\
\qquad \dfrac{f_4 - f_3}{x_4 - x_3} \equiv f_{34} \\
x_4 \quad f_4
\end{array} \qquad (7.25)
$$

Thus, we have achieved a form in which the coefficients are simple to evaluate, the final form gives rapid computation for many values of x by the nested-multiplication algorithm, and further interpolation points can be added to improve the accuracy of the scheme. The divided difference table shows that the addition of a further point simply introduces another diagonal line below the table, giving an additional coefficient a_{k+1} which does not alter the previous set. This is demonstrated in Example 7.1.

7.2.3 Iterative interpolation

The above scheme operates in two stages; firstly, the divided difference table is drawn up to calculate the coefficients, and secondly, nested multiplication is used to find the interpolated values. This procedure is most suitable when interpolated values are required at a large number of points, but in the case where only a few values are required the amount of computation can be reduced by using one of the iterative interpolation methods, such as those of Aitken and Neville.

The basis of these methods is to build up a table of polynomials, of similar form to the divided difference table, in which successive columns contain higher-order polynomials which fit the given data at a higher number of points as we move across the table. In calculating the interpolated values the polynomials themselves are not used. A particular value of x is inserted and a table of values corresponding to the various polynomials is produced. The degree of the polynomials increases across the table and, hopefully, the accuracy of the results increases also. Therefore, if the process is converging the results at the right-hand side of the table will eventually agree to the required accuracy. At this point the calculation is stopped. The table is built up by successive linear interpolations of previous polynomials according to the following formula.

$$
P_{r_1, r_2, \ldots, r_k, \iota, J}(x) = \frac{(x - x_I) P_{r_1 r_2, \ldots, r_k, J} - (x - x_J) P_{r_1, r_2, \ldots, r_k, \iota}}{(x_J - x_I)} \qquad (7.26)
$$

Initially, $P_0(x) = f_0$, $P_1(x) = f_1$, etc., and the first few entries of the Aitken table are

$$P_{01}(x) = \frac{(x - x_1)P_0(x) - (x - x_0)P_1(x)}{x_0 - x_1} \tag{7.27a}$$

$$P_{02}(x) = \frac{(x - x_2)P_0(x) - (x - x_0)P_2(x)}{x_0 - x_2} \tag{7.27b}$$

$$P_{012}(x) = \frac{(x - x_2)P_{01}(x) - (x - x_1)P_{02}(x)}{(x_1 - x_2)} \tag{7.27c}$$

The completed table has the following form.

x_0	f_0			
		$P_{01}(x)$		
x_1	f_1		$P_{012}(x)$	
		$P_{02}(x)$		$\cdot$
x_2	f_2		$\cdot$	
		$P_{03}(x)$	$\cdot$	
x_3	f_3		$\cdot$	
. . . .	. . . .	. . . .	. . . .	. . . .
x_k	f_k	$P_{0k}(x)$	$P_{01k}(x) \cdots P_{01 \ldots k}(x)$	

$$\tag{7.28}$$

If an extra function value is added to this table, the new row is calculated using only the values along the diagonal starting at f_0. It follows that only the values down this diagonal need to be stored, since all the other figures are intermediate values only.

The Neville table is based on the same formulae but uses different combinations of points to form the columns. This is shown in the table

x_0	f_0			
		$P_{01}(x)$		
x_1	f_1		$P_{012}(x)$	
		$P_{12}(x)$	$\cdot \cdot \cdot \cdot \cdot$	
x_2	f_2			$\cdot$
. . . .	. . . .	. . . .	. . . .	. . . .
x_k	f_k	$P_{k-1, k}(x)$	$P_{k-2, k-1, k}(x)$	$P_{012 \ldots k}(x)$

$$\tag{7.29}$$

When a new point is introduced into this table the calculations involve values from the previous row only, so that the rest of the table need not be stored. A numerical calculation using the Neville table is given in Example 7.2.

Either the iterative interpolation method or Newton's divided difference method are suitable for interpolation using a computer. In the iterative method a guide to the accuracy of the method is easily obtained from the degree of closeness of the successive members of the table, but each interpolation requires the calculation of the whole divided difference table. If many points are to be interpolated the Newton form is more suitable since only one evaluation of the table is performed. Each interpolation then requires only a nested multiplication which involves very little computation. However, preliminary investigations must be carried out to determine the number of interpolation points required to give sufficient accuracy.

7.3 Fitting by the method of least squares

The fundamental property of this method is that the sum of the squares of the errors is made as small as possible. Two situations arise, according to whether we approximate to a finite set of values or to a function defined over a range.

In the first case the error will be defined as the sum of the squares of the individual errors at each point and, in the second case an integral formulation is necessary. The latter formulation will be used to illustrate the theoretical basis of the method since most readers will be more familiar with integral calculus than summation properties.

In its general form the least-squares method is based on an approximating function which depends linearly on a set of parameters $a_0, a_1, \ldots, a_n$. The integral summation of the squares of the errors is given by

$$S = \int_a^b [f(x) - \phi(a_0, a_1, \ldots, a_n, x)]^2 \, dx \qquad (7.30)$$

Since we require S to be a minimum the first derivatives with respect to the various coefficients will be zero, i.e.,

$$\frac{\partial S}{\partial a_i} = 0 \qquad (7.31)$$

If the appropriate conditions hold for differentiation under the integral sign, then this gives $n + 1$ equations for the coefficients a_i, i.e.,

$$-2 \int_a^b \frac{\partial \phi}{\partial a_i} [f(x) - \phi(a_0, a_1, \ldots, a_n, x)] \, dx = 0, \qquad i = 0, 1, \ldots, n \qquad (7.32)$$

Since ϕ is a linear function of the coefficients the first terms of these equations are constant so that the equations can be written

$$\int_a^b \frac{\partial \phi}{\partial a_i} \phi(a_0, a_1, \ldots, a_n, x) \, dx = \int_a^b \frac{\partial \phi}{\partial a_i} f(x) \, dx, \qquad i = 0, 1, \ldots, n \qquad (7.33)$$

These equations are known as the normal equations.

As a simple illustration of the method, consider the case where a polynomial approximation is chosen, i.e.,

$$\phi(a_0, \ldots, a_n, x) = a_0 + a_1 x + \cdots + a_n x^n \qquad (7.34)$$

The normal equations then become

$$\int_a^b x^i [a_0 + a_1 x + \cdots + a_n x^n] \, dx = \int_a^b x^i f(x) \, dx, \qquad i = 0, 1, \ldots, n \qquad (7.35)$$

If these equations are written in matrix form

$$u_{00} a_0 + u_{01} a_1 + \cdots + u_{0n} a_n = b_0$$
$$\cdots\cdots\cdots\cdots\cdots\cdots\cdots\cdots\cdots\cdots\cdots$$
$$u_{n0} a_0 + u_{n1} a_1 + \cdots + u_{nn} a_n = b_n \qquad (7.36)$$

then the coefficients u_{ij} of the left-hand side matrix $\mathbf{U}$ are given by

$$u_{ij} = \int_a^b x^i . x^j \, dx, \qquad i, j = 0, 1, \ldots, n \qquad (7.37)$$

Unfortunately, these equations can be very ill-conditioned, and the solution of a least-squares problem directly by these methods is not recommended. Consider the example above with $a = 0$ and $b = +1$. The coefficients are given by

$$u_{ij} = 1/(i+j+1) \tag{7.38}$$

and

$$U = \begin{bmatrix} 1 & \frac{1}{2} & \frac{1}{3} & \frac{1}{4} & : \\ \frac{1}{2} & \frac{1}{3} & \frac{1}{4} & \frac{1}{5} & : \\ \frac{1}{3} & \frac{1}{4} & \frac{1}{5} & \frac{1}{6} & : \\ \frac{1}{4} & \frac{1}{5} & \frac{1}{6} & \frac{1}{7} & : \\ \cdots & \cdots & \cdots & \cdots & \end{bmatrix} \tag{7.39}$$

This matrix is the Hilbert matrix which it is well-known gives rise to ill-conditioned equations. This is demonstrated by the numerical inversion of one of the matrices in Example 7.6.

Ideally, we would like the normal equations to take a simple form which would give an efficient solution to the problem. The simplest form possible is the diagonal form, which would enable the coefficients $a_0, \ldots, a_n$ to be found directly by dividing by u_{ii} ($i = 0, 1, \ldots, n$). This form can be produced by taking advantage of the special properties of orthogonal functions.

The fundamental property of orthogonal functions is that for two different members of an orthogonal sequence $Q_k(x)$ ($k = 0, 1, 2, \ldots$)

$$\int_a^b w(x) \cdot Q_i(x) Q_j(x) \, dx = 0, \qquad i \neq j \tag{7.40}$$

provided the limits a and b and the weight function $w(x)$ are suitably chosen. In particular, if we choose $w(x) = 1$ and $a = -1, b = +1$ then the Legendre polynomials have the orthogonality property. Thus, instead of approximating by a sum of powers of x we use a sum of Legendre polynomials, i.e.,

$$\phi(a_0, a_1, \ldots, a_n, x) = a_0 \bar{P}_0(x) + a_1 \bar{P}_1(x) + \cdots + a_n \bar{P}_n(x). \tag{7.41}$$

The first few polynomials of this type are

$$\bar{P}_0(x) = 1, \qquad \bar{P}_1(x) = x, \qquad \bar{P}_2(x) = \tfrac{1}{2}(3x^2 - 1) \tag{7.42}$$

Corresponding to Eqns (7.36) we now have a set with only diagonal terms since

$$u_{ij} = \int_{-1}^{+1} \bar{P}_i(x) \bar{P}_j(x) \, dx = 0, \qquad \begin{matrix} i = 0, 1, \ldots, n \\ j \neq i \end{matrix} \tag{7.43}$$

The diagonal coefficients also take a particularly simple form.

$$u_{ii} = \int_{-1}^{+1} \bar{P}_i^2(x) \, dx = \frac{2}{2i+1} \tag{7.44}$$

so that the a_i are found directly

$$a_i = \frac{2i+1}{2} \int_{-1}^{+1} \bar{P}_i(x) f(x) \, dx, \qquad i = 0, 1, \ldots, n \tag{7.45}$$

These two alternative methods of calculation are shown in Example 7.3. If representation as a series of powers of x is required the expressions for $\bar{P}_i(x)$

can be substituted into the equations once the values of a_i have been found, and rearrangement will give the power series in x.

7.4 Orthogonal polynomials

In view of the importance of orthogonal polynomials in approximation problems some of the more important properties of these polynomials will be summarized here. We have a sequence of polynomials $Q_k(x)$ of degree k for $k \leqslant K$, for some fixed value of K.

7.4.1 Orthogonality relation

The polynomials $Q_k(x)$ are said to be orthogonal with respect to a weight function $w(x)$ over an interval $[a, b]$ if

$$\int_a^b w(x)Q_i(x)Q_j(x)\,dx = 0, \qquad i \neq j$$
$$w(x) \geqslant 0, \qquad a \leqslant x \leqslant b \tag{7.46}$$

If, in addition, the following condition is satisfied the polynomials form an orthonormal set.

$$\int_a^b w(x)Q_i^2(x)\,dx = 1 \quad \text{for all} \quad i \tag{7.47}$$

It should be noted that this definition implies that a polynomial $Q_i(x)$ is orthogonal to an arbitrary polynomial of degree $<i$. This follows because any polynomial can be expressed in terms of the set $Q_k(x)$ $(k = 0, 1, \ldots, i)$ because they are linearly independent, i.e.,

$$P_j(x) = \sum_{k=0}^{j} a_k Q_k(x), \qquad j < i \tag{7.48}$$

We see that

$$\int_a^b w(x)Q_i(x)P_j(x)\,dx = \sum_{k=0}^{j} a_k \int_a^b w(x)Q_i(x)Q_k(x)\,dx = 0 \tag{7.49}$$

All of the terms are zero by the orthogonality property so that $P_j(x)$ is orthogonal to $Q_i(x)$ $(j < i)$.

This orthogonality property does give a means for generating successive members of the sequence, although the recurrence property, which is discussed below, gives a more suitable method of practical implementation.

As an example, the coefficients of the second-degree Legendre polynomial can be found as follows. The orthogonality relation for the polynomial $ax^2 + bx + c$ gives

$$\int_{-1}^{+1} (ax^2 + bx + c).1.dx = 0, \qquad \frac{2a}{3} + 2c = 0 \tag{7.50}$$

and

$$\int_{-1}^{+1} (ax^2 + bx + c).x\,dx = 0, \qquad \frac{2b}{3} = 0 \tag{7.51}$$

Hence, the Legendre polynomial of degree 2 is $c(-3x^2+1)$. The orthogonality property holds for all values of c and the choice of this constant depends on a deeper knowledge of orthogonal function theory. The usual form of this Legendre polynomial is $\frac{1}{2}(3x^2-1)$ but if an orthonormal set is required the polynomial takes the form $(\sqrt{5}/2\sqrt{2})(3x^2-1)$.

7.4.2 Recurrence relation

It can be seen that the above method of finding the coefficients becomes impractical for high-order polynomials, since it involves solving a large set of simultaneous equations. Fortunately, another property of orthogonal functions gives a simpler method of generation which is suitable for computer use. All sets of orthogonal polynomials satisfy a recurrence relation of the following form.

$$Q_{n+1}(x)=(A_n+B_nx)Q_n(x)+C_nQ_{n-1}(x), \qquad n=1,2,\ldots \tag{7.52}$$

Thus, if the first two polynomials of the set are known, the third can be found from the above equation, and then the second and third used to generate the fourth member and so on. For the Legendre polynomials the recurrence relation is

$$\bar{P}_{n+1}(x)=\frac{2n+1}{n+1}\cdot x\bar{P}_n(x)-\frac{n}{n+1}\bar{P}_{n-1}(x) \tag{7.53}$$

The first two members of the sequence are $\bar{P}_0(x)=1, \bar{P}_1(x)=x$ and hence,

$$\bar{P}_2(x)=\tfrac{3}{2}\cdot x(x)-\tfrac{1}{2}\cdot 1=\tfrac{1}{2}(3x^2-1) \tag{7.54a}$$
$$\bar{P}_3(x)=\tfrac{5}{3}\cdot x(\tfrac{3}{2}x^2-\tfrac{1}{2})-\tfrac{2}{3}(x)=\tfrac{1}{2}(5x^3-3x) \tag{7.54b}$$

7.4.3 Discrete orthogonality

In the discussion so far, we have restricted our attention to the case where a continuous function is to be approximated, so that the integral orthogonality property is appropriate. If, however, only a discrete set of values of the function are available, then it is more appropriate to define the error in terms of these points only, i.e.,

$$S_m=\sum_{i=0}^{m}[\phi_n(x_i)-f_i]^2 \tag{7.55}$$

We must bear in mind that if $Q_n(x)$ is a polynomial of degree n, then it contains $n+1$ coefficients as variable parameters, and so a unique solution to the least-squares problem exists for $m\geqslant n+1$. The typical least-squares approximation usually has many more points m than the number of coefficients of the polynomial $n+1$.

If we again use an expansion in terms of orthogonal polynomials we have

$$\phi_n(x)=\sum_{k=0}^{n}a_kQ_k(x) \tag{7.56}$$

Using the conditions for a minimum

$$\frac{\partial S_m}{\partial a_r}=0 \tag{7.57}$$

the normal equations are generated

$$\sum_{k=0}^{n} a_k \sum_{i=0}^{m} Q_k(x_i)Q_r(x_i) = \sum_{i=0}^{m} Q_r(x_i)f_i, \qquad r=0,1,\ldots,n \qquad (7.58)$$

If the functions satisfy the discrete orthogonality property

$$\sum_{i=0}^{m} Q_k(x_i)Q_r(x_i) = 0, \qquad k \neq r \qquad (7.59)$$

then a set of diagonal equations has been produced which have the solutions

$$a_r = \frac{\displaystyle\sum_{i=0}^{m} Q_r(x_i)f_i}{\displaystyle\sum_{i=0}^{m} Q_r^2(x_i)} \qquad (7.60)$$

All the standard orthogonal polynomials have a set of points x_i $(i = 0, \ldots, m)$ over which a discrete orthogonality property holds. The polynomials which will be used to illustrate these properties are the Chebyshev polynomials, which are also of interest in connection with minimax approximation.

The Chebyshev polynomials are defined by

$$T_n(x) = \cos[n\cos^{-1}(x)], \qquad -1 \leqslant x \leqslant +1 \qquad (7.61)$$

and the zeros of this polynomial are given by

$$x_j = \cos\left[\frac{(2j+1)\pi}{2n}\right], \qquad j=0,1,\ldots,n-1 \qquad (7.62)$$

For a fixed value N all the Chebyshev polynomials $T_n(x)$ $(n < N)$ have the finite-orthogonality property with respect to the set of points

$$x_j = \cos\left(\frac{\pi j}{N}\right), \qquad j=0,1,\ldots,N \qquad (7.63)$$

The main problem associated with the use of discrete least-squares approximation is that we must be able to obtain the function values at any point x required by the formula, and this may not be possible. Also, there are difficulties if it is necessary to increase the degree of the approximating function. In the continuous case the calculation simply involves finding the quantities

$$a_k = \frac{\displaystyle\int_{a}^{b} Q_k(x)f(x)\,dx}{\displaystyle\int_{a}^{b} Q_k^2(x)\,dx} \qquad (7.64)$$

for further values of k, and this does not disturb the previous values.

In the discrete case the summation involves computing f_i at a new set of x values which, in general, are different for different values of m. One of the advantages of the Chebyshev formulation is that the x_j of order m also occur in the formulae of order $2m$, $4m$, etc., so that some of the f_i are known from previously calculated values.

Another point which requires some thought is the choice of the values n and m. In general, if the number of points m is increased this will give a closer approximation at the expense of more computation; a compromise must be made between accuracy and computational time. The choice of a suitable value depends on the particular problem and no general rule can be given.

Investigation into the choice of a suitable value for n gives considerable insight into the nature of least-squares approximations. When $n = m$ this produces the interpolation polynomial through the $n + 1$ points used. For values of $n < m$ the least-squares fit will not normally pass through the points, and the curve is subject to a smoothing process. This is particularly valuable when the method is applied to experimental results which give functional values together with experimental error. The small deviations due to the errors could result in a highly oscillatory polynomial which is essentially mapping the fluctuations due to the errors. Again a compromise must be sought between using a very low value of n which would smooth out even the fluctuations of the underlying functions and using too high a value of n when the experimental error would not be smoothed out. With curves which are expected to be fairly smooth the approximation can be achieved by splitting the range into several subranges. Then in each subrange a least-squares approximation based on a low-order polynomial, such as a cubic, is used.

7.4.4   Roots of the polynomials
If we have a set of polynomials $Q_k(x)$ $(k = 0, 1, \ldots)$ orthogonal over the interval $[a, b]$ then the roots x_j $(j = 1, 2, \ldots k)$ of $Q_k(x) = 0$ are all real and distinct and lie in the interval $a < x < b$.

This property is necessary when using the zeros of an orthogonal polynomial as the base points for a discrete least-squares approximation, as suggested in the last section. However, if there were coincident roots or complex roots it would not be possible to use these in the least-squares method.

7.4.5   Important orthogonal polynomials
We have already seen how the Legendre polynomials arise naturally in least-squares theory if the orthogonality property involves a weight function of unity. The apparent restriction of their applicability to the range $[-1, +1]$ can easily be overcome by a change of variable. If $x \epsilon [-1, +1]$ and $X \epsilon [a, b]$ then the equation $X = (x + 1)(b - a)/2 + a$ will effect a transformation to the standard range $[-1, +1]$. This change of variable is important in the theory of gaussian quadrature, since it means that any finite range can be considered by discussing the standard range $[-1, +1]$.

However, transformations such as the above cannot include the semi-infinite range $0, \infty$ or the infinite range $-\infty, +\infty$. The Laguerre polynomials are orthogonal over the range, $0, \infty$ and can be used for gaussian integration over this range. The Hermite polynomials are orthogonal over the range $-\infty, +\infty$.

Another set of orthogonal polynomials which are in frequent use are the Chebyshev polynomials which are orthogonal in the range $[-1, +1]$ with weight function $(1 - x^2)^{-1/2}$. The most significant property of these polynomials is their equal oscillation property. All the maxima and minima of the function occur in the interval $[-1, +1]$ and they all have the same absolute magnitude. It will be shown in the next section that this leads to the minimax property. Some of the properties of these polynomials are summarized in Table 7.1.

7.5 Chebyshev polynomials and minimax approximation

7.5.1 Definition of Chebyshev polynomials

The Chebyshev polynomials are a special case of the Jacobi polynomials with $p = q = \frac{1}{2}$. They are orthogonal over the interval $[-1, +1]$ with weight function $(1 - x^2)^{-1/2}$. The simplest introduction to these polynomials is through the cosine function, $\cos n\theta$ $(0 \leqslant \theta \leqslant \pi)$. The function, $\cos n\theta$, can be expanded using the trigonometric sum formulae, i.e.,

$$\cos n\theta = \cos [(n-1)\theta] \cos \theta - \sin [(n-1)\theta] \sin \theta \qquad (7.65)$$

and the terms of order $n - 1$ can then be further expanded and so on, until all the expressions contain θ only, and all multiples of θ have been expanded. The even powers of $\sin \theta$ which occur can then be replaced by $1 - \cos^2 \theta$ so that the expression now involves powers of $\cos \theta$ only.

It should be noted that $\cos n\theta$ is an even function and therefore can be expressed in terms of even functions only so it follows that the powers of $\sin \theta$ can be eliminated. For example,

$$\cos 2\theta = 2\cos^2 \theta - 1 \qquad (7.66a)$$
$$\cos 3\theta = \cos \theta [2\cos^2 \theta - 1] - \sin \theta . \sin 2\theta$$
$$= 2\cos^3 \theta - \cos \theta - 2\cos \theta [1 - \cos^2 \theta]$$
$$= 4\cos^3 \theta - 3\cos \theta \qquad (7.66b)$$

If the substitution $x = \cos \theta$ is made, the functions $\cos n\theta \equiv \cos (n\cos^{-1} x)$ are equivalent to polynomials of degree n for $-1 \leqslant x \leqslant +1$. Thus, a set of polynomials, known as the Chebyshev polynomials, can be defined by the expression

$$T_n(x) = \cos (n\cos^{-1} x), \qquad -1 \leqslant x \leqslant 1 \qquad (7.67)$$

This link with trigonometric functions is a useful aid for proving some of the properties of Chebyshev polynomials; a reformulation of the problem in terms of θ often leads to a simple proof.

For example, the integral orthogonality states that

$$\int_{-1}^{+1} \frac{T_r(x)T_s(x) \, dx}{(1 - x^2)^{1/2}} = 0, \qquad r \neq s \qquad (7.68)$$

Making the transformation $x = \cos \theta$ gives $dx = -\sin \theta \, d\theta$ and the integral becomes

$$\int_{-\pi}^{0} \cos r\theta . \cos s\theta \, d\theta = \frac{1}{2} \int_{-\pi}^{0} \cos [(r+s)\theta] + \cos [(r-s)\theta] \, d\theta \qquad (7.69)$$

The integral of $\cos n\theta$ where n is an integer over the range $-\pi \leqslant \theta \leqslant 0$ is zero unless $n = 0$ so that both parts of the integral become zero except when $r = s$. Hence, the orthogonality condition is proved.

Table 7.1

	$w(x)$	a	b	First term	Second term	Third term	Recurrence relation
Jacobi	$(1-x)^{-p}(1+x)^{-q}$ $p<1, q<1$	-1	$+1$				
Legendre	1 $p=0, q=0$	-1	$+1$	1	x	$\frac{1}{2}(3x^2-1)$	$P_{n+1}(x)=\dfrac{2n+1}{n+1}\cdot xP_n(x)-\dfrac{n}{n+1}P_{n-1}(x)$
Chebyshev	$(1-x^2)^{-1/2}$ $p=\frac{1}{2}, q=\frac{1}{2}$	-1	$+1$	1	x	$2x^2-1$	$T_{n+1}(x)=2xT_n(x)-T_{n-1}(x)$
Laguerre	$e^{-\alpha x}$	0	∞	1	$1-x$	$2-4x+x^2$	$L_{r+1}(x)=(1+2r-x)L_r(x)-r^2L_{r-1}(x)$
Hermite	$e^{-\alpha^2 x^2}$	$-\infty$	$+\infty$	1	$2x$	$4x^2-2$	$H_{r+1}(x)=2xH_r(x)-2rH_{r-1}(x)$

The recurrence relation for the Chebyshev polynomials can be derived in the same way.

$$T_{n+1}(x) \equiv \cos(n+1)\theta = \cos n\theta \cos\theta - \sin n\theta \sin\theta$$
$$T_{n-1}(x) \equiv \cos(n-1)\theta = \cos n\theta \cos\theta + \sin n\theta \sin\theta \qquad (7.70)$$

Hence,

$$T_{n+1}(x) + T_{n-1}(x) = 2x\cos n\theta \quad \text{or} \quad T_{n+1}(x) = 2xT_n(x) - T_{n-1}(x) \quad (7.71)$$

7.5.2 The minimax property

The most important property of Chebyshev polynomials is the minimax property which is a consequence of the equi-oscillation property.

The Chebyshev polynomial $T_n(x)/2^{n-1}$ has the least maximum modulus, in the range $[-1, +1]$, of all polynomials of degree n with highest coefficient unity. This means that all other polynomials move further away from zero, at some point in the range, than the normalized Chebyshev polynomial which oscillates between $\pm(\frac{1}{2})^{n-1}$.

This property is easily proved from the properties of the cosine function. $\cos n\theta$ has maxima and minima at the points

$$\theta_j = \frac{j\pi}{n}, \qquad j = 0, 1, \ldots, n \qquad (7.72)$$

where it takes the values ± 1 so that the Chebyshev polynomial takes the values $\pm(\frac{1}{2})^{n-1}$ alternately. Assume that there is a polynomial $p_n(x)$, with highest coefficient unity, whose values are less in modulus than $(\frac{1}{2})^{n-1}$ at all points in the range $-1 \le x \le +1$. We will then show that this assumption leads to a contradiction.

Form the function

$$\phi_{n-1}(x) = p_n(x) - \frac{T_n(x)}{2^{n-1}} \qquad (7.73)$$

The subscript $n-1$ is used to emphasize that the two terms in x^n cancel out, since both polynomials have been normalized to make their leading coefficient unity.

At each of the points $\theta_j = j\pi/n$, where j is even, we know that $\phi_{n-1}(x)$ is negative, since the value of $-T_n(x)/2^{n-1}$ is $-1/2^{n-1}$ which has larger modulus than $p_n(x)$ by the assumption. Similarly, at the points θ_j where j is odd, the value of $\phi_{n-1}(x)$ must be positive since $-T_n(x)/2^{n-1}$ takes its maximum positive value of $(\frac{1}{2})^{n-1}$.

Hence, at the $n+1$ points θ_j the polynomial $\phi_{n-1}(x)$ alternates between positive and negative values. Since the function is continuous there must be n real and distinct roots between these points. This means that $\phi_{n-1}(x)$ is a polynomial of degree $n-1$ with n values equal to zero. This is not possible so the original assumption must be unsound.

It should be noted that the conclusion was based on the fact that the polynomial reached its maximum and minimum distances from zero alternatively $n+1$ times. The more general problem of minimax or Chebyshev approximation is to find a function which has this equi-oscillation property with respect to some other function rather than zero. We have shown that a Chebyshev polynomial is the best approximation to zero in the minimax sense. We may, however, require a best approximation to other functions such as a sine or a logarithmic function. This topic is outside the scope of this text but it will be shown that useful approximations can be achieved based on the Chebyshev polynomials, although not the best approximation in the minimax sense.

7.5.3 Economization of polynomials

The above property of Chebyshev polynomials can be used to find the best polynomial approximation of degree $n-1$ to a given polynomial of degree n. If the given polynomial is

$$p_n(x) = a_0 + a_1x + \cdots + a_nx^n, \qquad -1 \leqslant x \leqslant +1 \tag{7.74}$$

then we form the polynomial

$$q_{n-1}(x) = p_n(x) - \frac{a_n}{2^{n-1}} T_n(x) \tag{7.75}$$

It can be seen that $q_{n-1}(x)$ is a polynomial with maximum degree $n-1$, since the coefficient of x^n has been chosen to be zero. The difference between the two polynomials $q_{n-1}(x)$ and $p_n(x)$ is a multiple of a Chebyshev polynomial, and therefore deviates from zero by the least amount in the interval $-1 \leqslant x \leqslant +1$. The process of economization relies on removing the term of highest degree by subtracting a multiple of the appropriate Chebyshev polynomial.

When a new polynomial $q_{n-1}(x)$ has been found, it is then possible to form a best approximation to this by the same method.

$$q_{n-2}(x) = q_{n-1}(x) - \frac{b_{n-1}}{2^{n-2}} T_{n-1}(x) \tag{7.76}$$

Each member of the sequence would be a best approximation to the previous member, but it must be realized that the approximation to $p_n(x)$ will only be a best approximation, in the minimax sense, for the polynomial $q_{n-1}(x)$. For example, the error of the second approximation is given by

$$q_{n-2}(x) - p_n(x) = -\frac{b_{n-1}}{2^{n-2}} T_{n-1}(x) - \frac{a_n}{2^{n-1}} T_n(x) \tag{7.77}$$

and the sum of two Chebyshev polynomials will not possess the minimax property. However, it is found in practice that this process of economization gives a good approximation of lower degree.

The maximum error possible in the approximation is easily found, since the $T_r(x)$ have maxima and minima of ± 1. Thus, for $q_{n-2}(x)$ we have an error which has a maximum modulus less than

$$\left| \frac{b_{n-1}}{2^{n-2}} \right| + \left| \frac{a_n}{2^{n-1}} \right|$$

Consider the expansion

$$\log(1+x) = x - \frac{x^2}{2} + \frac{x^3}{3} - \frac{x^4}{4} + \frac{x^5}{5} \cdots, \qquad -1 < x \leqslant +1 \tag{7.78}$$

This expansion converges very slowly; indeed, this is often a problem with Taylor series expansions. Thus, the deletion of the last two terms from the above approximation introduces an error which has a maximum modulus less than $\frac{1}{4} + \frac{1}{5} = \frac{9}{20}$.

The above approximation can be expressed in terms of Chebyshev polynomials by using the following expressions which are easily obtained from the Chebyshev polynomial formulae

$$1 = T_0(x) \qquad\qquad\qquad x = T_1(x)$$
$$x^2 = \tfrac{1}{2}[T_0(x) + T_2(x)] \qquad\qquad x^3 = \tfrac{1}{4}[3T_1(x) + T_3(x)] \qquad (7.79)$$
$$x^4 = \tfrac{1}{8}[3T_0(x) + 4T_2(x) + T_4(x)] \quad x^5 = \tfrac{1}{16}[10T_1(x) + 5T_3(x) + T_5(x)]$$

$$\log(1+x) =$$
$$-\tfrac{11}{32}T_0(x) + \tfrac{11}{8}T_1(x) - \tfrac{3}{8}T_2(x) + \tfrac{7}{48}T_3(x) - \tfrac{1}{32}T_4(x) + \tfrac{1}{80}T_5(x) - \cdots \quad (7.80)$$

The deletion of the last two terms of this approximation introduces an error of maximum modulus less than $\frac{1}{32} + \frac{1}{80} = \frac{7}{160}$, which is a substantial improvement on the previous result. The reader will appreciate that this example was chosen for simplicity, and the numbers involved are unrealistic, but the principle of economization can be used most effectively to give a lower-degree polynomial approximation with a relatively small loss of accuracy. (See also Example 7.7.)

7.5.4   Chebyshev series expansion

As an alternative to expansion by Taylor series followed by economization, it may be possible to expand directly as a Chebyshev expansion. Any function $f(x)$ which is continuous and has a finite number of maxima and minima in the range $[-1, +1]$ can be expanded in this way, i.e.,

$$f(x) = \sum_{k=0}^{\infty}{}' a_k T_k(x) \qquad\qquad (7.81)$$

The prime signifies that the first term is $a_0/2$, which simplifies the definitions of the coefficients a_r in the same way as for Fourier coefficients.

By the standard orthogonality property the coefficients can be evaluated. Multiply Eqn (7.81) by $(1 - x^2)^{-1/2} \cdot T_s(x)$ and then integrate between the limits -1 and $+1$. All terms except one on the right-hand side become zero which gives

$$a_s = \frac{2}{\pi} \int_{-1}^{+1} \frac{T_s(x) \cdot f(x)\, dx}{(1 - x^2)^{1/2}} \qquad\qquad (7.82)$$

It would, of course, be necessary to evaluate this integral by approximate means if we were using computer methods and, therefore, it is preferable to have a scheme based on a finite series, using finite orthogonality properties. Assume that the values of $f(x)$ are known at the points $x_j = \cos(\pi j/n)$ $(j = 0, 1, \ldots, n)$ and choose a function

$$\phi(x) = \sum_{k=0}^{n}{}'' b_k T_k(x) \qquad\qquad (7.83)$$

so that the values coincide at the points x_j $(j = 0, 1, \ldots, n)$.

$$f(x_j) = \phi(x_j) = \sum_{k=0}^{n}{}'' b_k T_k(x_j), \qquad i = 0, 1, \ldots, n \tag{7.84}$$

The Σ'' indicates that the first and last terms have coefficients $b_0/2$ and $b_n/2$ respectively.

The finite orthogonality property is defined by

$$\sum_{j=0}^{n}{}'' T_r(x_j)T_s(x_j) = \begin{cases} n & r = s = 0, n \\ n/2 & r = s \neq 0 \\ 0 & r \neq s \end{cases} \tag{7.85}$$

The coefficients b_r can be found by multiplying each of Eqns (7.84) by the appropriate value $T_s(x_j)$ $(j = 0, 1, \ldots, n)$ and summing the equations. The orthogonality property means that all terms except one on the right-hand side become zero giving

$$b = \frac{2}{n}\sum_{j=0}^{n}{}'' f(x_j) \cdot T_s(x_j) \tag{7.86}$$

It is clear that this expression can be directly calculated on a computer. A simple Chebyshev series expansion is given in Example 7.4.

Naturally, it is of interest to know how close is the connection between the coefficients defining the finite Chebyshev expansion, given by $\phi(x)$, and the coefficients of the infinite series. If the infinite series expression for $f(x)$ is inserted in Eqn (7.86) we have

$$b_s = \frac{2}{n}\sum_{j=0}^{n}{}'' \sum_{k=0}^{\infty}{}' a_k T_k(x_j)T_s(x_j) \tag{7.87}$$

Using the definition of the Chebyshev polynomial $T_k(x) = \cos k\theta$, $\cos \theta = x$ we can use the properties of the cosine function to show that all the terms $k = 2n \cdot M \pm s$ will satisfy the finite orthogonality property (7.85), if M is a positive integer ≥ 1. This follows since

$$\cos \frac{k \cdot \pi j}{n} = \cos \frac{k\pi s}{n}$$

for values of k such that

$$\frac{k\pi j}{n} = 2\pi M' \pm \frac{k\pi s}{n} \quad \text{or} \quad k = \frac{2M'n}{j} \pm s \tag{7.88}$$

Since k must be integer we are restricted to those values of M' which are multiples of j, giving $k = 2Mn \pm s$.

Thus, only these values of k will lead to nonzero values on the right-hand side, and this gives

$$b_s = a_s + a_{2n-s} + a_{2n+s} + a_{4n-s} + a_{4n+s} + \cdots \tag{7.89}$$

If the series has good convergence properties these higher terms will soon be negligible, and the b_s will be a good approximation to the coefficients of the infinite series.

7.5.5 Evaluation of Chebyshev series

It has been pointed out in Chapter 3 that the evaluation of polynomials should be carried out using the nested-multiplication algorithm. So one possible way of evaluating a finite Chebyshev series (7.81) would be to calculate the coefficients of the various powers of x, and use this algorithm. However, there is an algorithm which can be used on the Chebyshev series directly, which is very similar to the nested-multiplication algorithm. This method is preferable since it avoids the rearrangement into a power series in x. A numerical example is given by Example 7.5.

Let

$$c_{n+1} = c_{n+2} = 0$$

and use the recurrence relation

$$c_r = 2xc_{r+1} - c_{r+2} + b_r, \qquad r = n, n-1, \ldots, 0 \tag{7.90}$$

The value of the Chebyshev series (7.81) at the point x is given by

$$\phi(x) = \tfrac{1}{2}[c_0 - c_2] \tag{7.91}$$

7.5.6 Other properties

The Chebyshev equi-oscillation property has an interesting application when choosing the points to be used as the basis for an interpolation formula. It was shown in Section 7.2.2 that the error of the interpolation formula using the points $x_0, \ldots, x_n$, is given by

$$(x - x_0)(x - x_1) \ldots (x - x_n)\frac{f^{(n+1)}(\zeta)}{(n+1)!} \tag{7.92}$$

where ζ is some point in the interval in which the $n+1$ points, x_i, lie. Naturally, we require that this error term be as small as possible. Although little can be done to minimize the derivative term it is certainly possible to give the product terms the minimum oscillation by choosing the points x_i $(i = 0, 1, \ldots, n)$ to be the zeros of the Chebyshev polynomial $T_{n+1}(x)$.

It is also of interest that the Chebyshev polynomials are used in gaussian integration formulae. They lead to a formula which has equal coefficients, which slightly reduces the computation required. Also, the presence of the weight factor, $(1 - x^2)^{-1/2}$, means that certain integrals with singularities in the integrand can still be evaluated numerically. Further details are given in Section 8.3.

Bibliographic notes

The literature on approximation is very extensive, and it is only possible to mention a few texts for further reading.

A useful discussion with several examples is given in Fox and Mayers (1968) which covers polynomial and Chebyshev approximation and finite-difference interpolation. The books of Lanczos (1957) and Hamming (1962) contain extensive material on approximation by infinite and finite Fourier series. The reader is particularly recommended to the book by Lanczos, which gives useful insight into methods as well as mathematical discussion; many ideas are suggested for future development. The book by Hamming also has a good coverage of polynomial and least-squares approximation.

A treatment of the topics at a more formal analytical level is provided by

Ralston (1965) and Isaacson and Keller (1966). The reader concerned with the mathematical theorems associated with these methods will find the latter text particularly useful. More detailed works on approximation suitable for the serious student in this field are Davis (1964) and Rice (1964).

Worked examples

1 A divided difference table for the following function values is given below.

x	$f(x)$			
1·6	0·6250			
2·9	0·3448	−0·2155		
3·7	0·2703	−0·0931	+0·0582	
4·8	0·2083	−0·0563	+0·0193	−0·0121

The interpolation formula derived from a divided difference table has the form

$$P_3(x) = a_0 + a_1(x - x_0) + a_2(x - x_0)(x - x_1) + a_3(x - x_0)(x - x_1)(x - x_2)$$
$$= ((a_3(x - x_2) + a_2)(x - x_1) + a_1)(x - x_0) + a_0$$

where the a_i ($i = 0, 1, 2, 3$) are the respective divided differences, i.e., $a_0 = 0.6250$, $a_1 = -0.2155$, $a_2 - 0.0582$, and $a_3 = -0.0121$. The second form of the interpolation polynomial is in the nested-multiplication form which is convenient for numerical calculations.

The interpolated values at $x = 2.0$ and $x = 4.5$ are found by nested multiplication to be 0·5105 and 0·2254 respectively.

If higher accuracy is required a further point can be added which gives an additional row to the divided difference table. Let the new point be $x = 4.0$, $f(x) = 0.25$. The new row is

4·0   0·25   −0·0521   0·0140   −0·0048   0·0030

Thus, the additional coefficient $a_4 = 0.0030$ and the nested multiplication gives interpolated values at the points $x = 2.0$ and $x = 4.5$ of 0·5058 and 0·2221, respectively. The correct values of the function are 0·5 and 0·2222.

2 The Neville table for the above function values at the point $x = 2.0$ is shown below. The additional row of values for $x = 4.0$ can easily be added after the main table has been calculated.

x	$f(x)$			
1·6	0·6250			
2·9	0·3448	0·5387		
3·7	0·2703	0·5574	0·5176	
4·8	0·2083	0·5729	0·5224	0·5101

The first entry in the table is found from the formula

$$\frac{(x - x_0)f_1 - (x - x_1)f_0}{x_1 - x_0} = \frac{0.4 \times 0.3448 + 0.9 \times 0.6250}{2.9 - 1.6}$$
$$= 0.5387$$

The first entry in the third column is calculated from the formula

$$\frac{(x - x_0)f_{12} - (x - x_2)f_{01}}{x_2 - x_0} = \frac{0.4 \times 0.5574 + 1.7 \times 0.5387}{3.7 - 1.6}$$
$$= 0.5176$$

The addition of the point $x = 4.0, f(x) = 0.25$ gives the following additional row.

4·0   0·25   0·5625   0·5192   0·5086   0·5047

Thus, the most accurate value in the first table is 0·5101 and the extra row gives the value 0·5047.

3 The following table of values is to be approximated by a cubic using the least-squares method.

x	-3	-2	-1	0	1	2	3
$f(x)$	-27	-16	-9	1	8	18	26

The error term S is given by

$$S = \sum_{i=0}^{6} [a_0 + a_1 x_i + a_2 x_i^2 + a_3 x_i^3 - f_i]^2$$

The various coefficients of the matrix which is used to find the coefficients a_i $(i = 0, 1, \ldots, 3)$ are

$$\begin{matrix} u_0 & u_1 & u_2 & u_3 \\ u_1 & u_2 & u_3 & u_4 \\ u_2 & u_3 & u_4 & u_5 \\ u_3 & u_4 & u_5 & u_6 \end{matrix}$$

where

$$u_r = \sum_{i=0}^{6} x_i^r$$

The right-hand side terms are $b_r = \sum_{i=0}^{6} x_i^r f_i$. Evaluating the various coefficients gives the equations

$$\begin{bmatrix} 7 & 0 & 28 & 0 \\ 0 & 28 & 0 & 196 \\ 28 & 0 & 196 & 0 \\ 0 & 196 & 0 & 1588 \end{bmatrix} \begin{bmatrix} a_0 \\ a_1 \\ a_2 \\ a_3 \end{bmatrix} = \begin{bmatrix} 1 \\ 244 \\ -2 \\ 1720 \end{bmatrix}$$

Subtracting four times the first equation, from the third, gives

$$(196 - 112)a_2 = -6$$

Therefore,

$$a_2 = -\tfrac{1}{14}, \qquad a_0 = \tfrac{3}{7}$$

Subtracting seven times the second equation from the fourth equation, gives

$$(1588 - 1372)a_3 = 1720 - 1708$$

Therefore,

$$a_3 = \tfrac{1}{18}, \qquad a_1 = \tfrac{1}{28}(244 - \tfrac{196}{18}) = \tfrac{1049}{126}$$

Therefore, the required least-squares polynomial is

$$P_3(x) = \frac{3}{7} + \frac{1049}{126}x - \frac{x^2}{14} + \frac{x^3}{18}$$

This problem was made very much easier by the simple form of the matrix. In general the solution must be found by full gaussian elimination.

The problem can also be solved by using the polynomials which are orthogonal over this particular set of points, i.e.,

$$Q_0(x) = 1, \qquad Q_1(x) = x, \qquad Q_2(x) = x^2 - 4, \qquad Q_3(x) = x^3 - 7x$$

Hence, letting $P_3(x) = a_0 Q_0(x) + a_1 Q_1(x) + a_2 Q_2(x) + a_3 Q_3(x)$, the equations become

$$7a_0 \qquad\qquad\qquad = -27 - 16 - 9 + 1 + 8 + 18 + 26 = +1$$
$$28a_1 \qquad\qquad = 81 + 32 + 9 + 8 + 36 + 78 = 244$$
$$84a_2 \qquad\qquad = -6$$
$$216a_3 = 12$$

$$a_0 = \tfrac{1}{7}, \qquad a_1 = \tfrac{61}{7}, \qquad a_2 = -\tfrac{1}{14}, \qquad a_3 = \tfrac{1}{18}$$

The required polynomial is, therefore,

$$P_3(x) = \tfrac{1}{7} + \tfrac{61}{7}x - \frac{(x^2 - 4)}{14} + \tfrac{1}{18}(x^3 - 7x)$$

It is left to the reader to verify that these two solutions are the same.

The diagonal equations can be solved directly. With a large number of equations, if full gaussian elimination for the ordinary polynomial method then the orthogonal polynomial method results in a simpler solution.

4 Find a finite Chebyshev series expansion of four terms which approximates the function e^x in the range $[-1, +1]$.

We wish to find the coefficients in the expansion

$$\phi(x) = \frac{b_0}{2}T_0(x) + b_1 T_1(x) + b_2 T_2(x) + \frac{b_3}{2}T_3(x)$$

given the values at the appropriate points for finite orthogonality, namely, $x_j = \cos(\pi j/n), j = 0, 1, 2, 3$.

The values of the function and the various Chebyshev polynomials at these points are

x	$f(x)$	$T_0(x)$	$T_1(x)$	$T_2(x)$	$T_3(x)$
1	2·718 282	1	1	1	$+1$
$\tfrac{1}{2}$	1·648 721	1	$\tfrac{1}{2}$	$-\tfrac{1}{2}$	-1
$-\tfrac{1}{2}$	0·606 530	1	$-\tfrac{1}{2}$	$-\tfrac{1}{2}$	$+1$
-1	0·367 879	1	-1	1	-1

The finite orthogonality relationship is

$$\sum_{i=0}^{3}{}'' T_r(x_j)T_s(x_j) = \begin{cases} 0 & r \neq s \\ \dfrac{2}{n} & r = s = 1, 2 \\ n & r = s = 0, 3 \end{cases}$$

where the double prime signifies that the first and last term are multiplied by $\tfrac{1}{2}$. Thus, the formula for the coefficients is

$$b_s = \frac{2}{n}\sum_{j=0}^{n}{}'' f(x_j)T_s(x_j)$$

The first and last coefficients, b_0 and b_3, have a weight factor of $\tfrac{1}{2}$ so that a single formula can be used to define all the coefficients. Thus,

$$b_0 = \frac{2}{3}\left[\frac{e}{2} + e^{1/2} + e^{-1/2} + \frac{e^{-1}}{2}\right] \qquad = 2·532\,221$$

$$b_1 = \frac{2}{3}\left[\frac{e}{2} + \frac{1}{2}e^{-1/2} - \frac{1}{2}e^{-1/2} - \frac{e^{-1}}{2}\right] = 1·130\,865$$

$$b_2 = \frac{2}{3}\left[\frac{e}{2} - \frac{1}{2}e^{1/2} - \frac{1}{2}e^{-1/2} + \frac{e^{-1}}{2}\right] = 0·276\,970$$

$$b_3 = \frac{2}{3}\left[\frac{e}{2} - e^{-1/2} + e^{-1/2} - \frac{e^{-1}}{2}\right] = 0·088\,674$$

Hence,

$$\phi(x) = 1·266\,110 + 1·130\,865T_1(x) + 0·276\,970T_2(x) + 0·044\,337T_3(x)$$

The nested-multiplication method of evaluating the Chebyshev series can be used. The formulae given in Eqns (7.90) and (7.91) refer to a Chebyshev expansion with $b_0/2$ as the first term but the last term is b_n not $b_n/2$. Thus, the value of b_3 which must be used is 0·044 337 but the value of b_0 is 2·532 222 1 in both cases.

As an example at the point $x = 1$

$$c_3 = 0 - 0 + 0·044 337$$
$$c_2 = 2 \times 0·044 337 - 0 + 0·276 970 = 0·365 644$$
$$c_1 = 2 \times 0·365 644 - 0·044 337 + 1·130 865 = 1·817 816$$
$$c_0 = 2 \times 1·817 816 - 0·365 644 + 2·532 221 = 5·802 209$$

Answer $= \frac{1}{2}[5·802 209 - 0·365 644]$
$= 2·718 282$

5 Evaluate the above Chebyshev series at the points $x = 0$, 0·25, and $-0·25$ and compare the answers with the true values, i.e., $x = 1$, 1·284 025, and 0·778 801.

$x = 0$

$$c_3 = 0 - 0 + 0·044 337$$
$$c_2 = 0 - 0 + 0·276 970$$
$$c_1 = 0 - 0·044 337 + 1·130 865 = 1·086 528$$
$$c_0 = 0 - 0·276 970 + 2·532 221 = 2·255 251$$

Answer $= \frac{1}{2}[c_0 - c_2] = \dfrac{1·978 281}{2}$
$= 0·989 140$

$x = 0·25$

$$c_3 = 0 - 0 + 0·044 337$$
$$c_2 = 0·044 337 \times \tfrac{1}{2} - 0 + 0·276 970 = 0·299 138$$
$$c_1 = 0·299 138 \times \tfrac{1}{2} - 0·044 337 + 1·130 865 = 1·236 097$$
$$c_0 = 1·236 097 \times \tfrac{1}{2} - 0·299 138 + 2·532 221 = 2·851 131$$

Answer $= \frac{1}{2}[2·851 131 - 0·299 138] = 1·275 996$

$x = -0·25$

$$c_3 = 0 - 0 + 0·044 337$$
$$c_2 = -0·044 337 \times \tfrac{1}{2} - 0 + 0·276 970 = 0·254 802$$
$$c_1 = -0·254 802 \times \tfrac{1}{2} - 0·044 337 + 1·130 865 = 0·959 127$$
$$c_0 = -0·959 127 \times \tfrac{1}{2} - 0·254 802 + 2·532 221 = 1·797 855$$

Answer $= \frac{1}{2}[1·797 855 - 0·254 802] = 0·771 527$

6 The Hilbert matrices are ill-conditioned for the solution of linear simultaneous equations. This is shown by calculating the inverse of a three by three Hilbert matrix using six significant figures which gives a fair degree of accuracy and using three significant figures which gives very inaccurate results. The determinant of the matrix is approximately 0·000 463 which indicates the ill-conditioned nature of the matrix. The calculation tableau using gaussian elimination to find the inverse to six significant figures is

1·000 000	0·500 000	0·333 333	1·000 00	0·000 00	0·000 00
0·500 000	0·333 333	0·250 000	0·000 00	1·000 00	0·000 00
0·333 333	0·250 000	0·200 000	0·000 00	0·000 00	1·000 00
	0·083 333 3	0·083 333 3	$-0·500 000$	1·000 00	0·000 00
	0·083 333 3	0·088 888 9	$-0·333 333$	0·000 00	1·000 00
		0·005 555 6	0·166 667	$-1·000 00$	1·000 00

Back substitution

$$R_3 = 29·9977 \quad -179·986 \quad 179·986$$
$$R_2 = -35·9977 \quad 191·986 \quad -179·986$$
$$R_1 = 8·999 62 \quad -35·9977 \quad 29·9977$$

The results can be checked by multiplication.

$$\begin{bmatrix} 8\cdot999\,62 & -35\cdot9977 & 29\cdot9977 \\ -35\cdot9977 & 191\cdot986 & -179\cdot986 \\ 29\cdot9977 & -179\cdot986 & 179\cdot986 \end{bmatrix} \begin{bmatrix} 1\cdot000\,00 & 0\cdot500\,000 & 0\cdot333\,333 \\ 0\cdot500\,000 & 0\cdot333\,333 & 0\cdot250\,000 \\ 0\cdot333\,333 & 0\cdot250\,000 & 0\cdot200\,000 \end{bmatrix}$$

$$= \begin{bmatrix} 0\cdot999\,993 & 0\cdot000\,14 & -0\cdot000\,015 \\ 0\cdot000\,027 & 0\cdot999\,919 & 0\cdot000\,079 \\ -0\cdot000\,027 & 0\cdot000\,077 & 0\cdot999\,923 \end{bmatrix}$$

The same tableau to three significant figures gives answers which have very poor accuracy.

1·00	0·500	0·333	1·00	0·00	0·00
0·500	0·333	0·250	0·00	1·00	0·00
0·333	0·250	0·200	0·00	0·00	1·00
	0·0833	0·0833	−0·500	1·00	0·00
	0·0833	0·0889	−0·333	0·00	1·00
		0·0056	0·167	−1·00	1·00

Back substitution

$$\begin{array}{rrrr} R_3 = & 29\cdot8 & -179 & +179 \\ R_2 = & -35\cdot8 & 191 & -179 \\ R_1 = & 8\cdot98 & -35\cdot9 & 29\cdot9 \end{array}$$

The results are checked by multiplication.

$$\begin{bmatrix} 8\cdot98 & -35\cdot9 & 29\cdot9 \\ -35\cdot8 & 191 & -179 \\ 29\cdot8 & -179 & 179 \end{bmatrix} \begin{bmatrix} 1\cdot00 & 0\cdot500 & 0\cdot333 \\ 0\cdot500 & 0\cdot333 & 0\cdot250 \\ 0\cdot333 & 0\cdot250 & 0\cdot200 \end{bmatrix}$$

$$= \begin{bmatrix} 0\cdot987 & 0\cdot0103 & -0\cdot004\,66 \\ 0\cdot0930 & 0\cdot953 & 0\cdot0286 \\ -0\cdot0930 & 0\cdot0430 & 0\cdot973 \end{bmatrix}$$

If the above calculations are carried out to three decimal places instead of three significant figures then the inverse, and the product of the inverse and the original matrix, are given respectively by

8·661	−33·858	27·834		1·001	0·015	−0·014
−33·857	178·715	−166·667		0·001	0·916	0·072
27·833	−166·667	166·667		−0·001	0·084	0·934

7 The Taylor Series for e^x up to the fifth power of x in the range $-1 \leqslant x \leqslant 1$ has a maximum error of $0\cdot003\,775$. By expressing the powers of x in terms of Chebyshev polynomials show that if the last two terms of the Chebyshev expansion are omitted the additional error introduced is less than $0\cdot006$.

The truncated Taylor series is

$$e^x = 1 + x + \frac{x^2}{2} + \frac{x^3}{6} + \frac{x^4}{24} + \frac{x^5}{120}$$

with a maximum error of $e^{1\cdot0}/720 = 0\cdot003\,775$. The following relations are used for the various powers of x.

$$1 = T_0(x), \qquad x = T_1(x), \qquad x^2 = \tfrac{1}{2}[T_2(x) + T_0(x)]$$
$$x^3 = \tfrac{1}{4}[T_3(x) + 3T_1(x)], \qquad x^4 = \tfrac{1}{8}[T_4(x) + 4T(x) + 3T_0(x)]$$
$$x^5 = \tfrac{1}{16}[T_5(x) + 5T_3(x) + 10T_1(x)]$$

Hence,

$$\begin{aligned} e^x &= T_0(x) + T_1(x) + \tfrac{1}{4}[T_2(x) + T_0(x)] + \tfrac{1}{24}[T_3(x) + 3T_1(x)] \\ &\quad + \tfrac{1}{192}[T_4(x) + 4T_2(x) + 3T_0(x)] + \tfrac{1}{1920}[T_5(x) + 5T_3(x) + 10T_1(x)] \\ &= \tfrac{443}{192}T_0(x) + \tfrac{217}{182}T_1(x) + \tfrac{13}{48}T_2(x) + \tfrac{17}{384}T_3(x) + \tfrac{1}{192}T_4(x) + \tfrac{1}{1920}T_5(x) \end{aligned}$$

Since the Chebyshev polynomials oscillate between ± 1, the maximum modulus of the sum of the last two terms is $0{\cdot}005\,208 + 0{\cdot}000\,520\,8 \approx 0{\cdot}005\,728$.

Problems

1 Use lagrangian interpolation to find function values at the points $1{\cdot}2$, $1{\cdot}9$, and $2{\cdot}1$, given the following table of values.

x	$1{\cdot}0$	$1{\cdot}5$	$2{\cdot}6$	$2{\cdot}8$	$3{\cdot}0$
$f(x)$	$2{\cdot}7183$	$4{\cdot}4817$	$13{\cdot}464$	$16{\cdot}445$	$20{\cdot}086$

2 Use lagrangian interpolation to find function values at the points $x = 2{\cdot}0$ and $4{\cdot}5$, given the following table of values.

x	$1{\cdot}6$	$2{\cdot}9$	$3{\cdot}7$	$4{\cdot}8$
$f(x)$	$0{\cdot}6250$	$0{\cdot}3448$	$0{\cdot}2703$	$0{\cdot}2083$

3 Given the Chebyshev polynomials $T_0(x) = 1$ and $T_1(x) = x$ and the recurrence relation

$$T_{r+1}(x) + T_{r-1}(x) = 2xT_r(x), \qquad r = 1, 2, \ldots,$$

find $T_2(x)$, $T_3(x)$, and $T_4(x)$. Use these results to find expressions for the powers of x up to x^4 in terms of $T_r(x)$ $(r = 0, 1, \ldots, 4)$.

4 Using the approximation to $\log(1 + x)$ in the interval $[0, 1]$ given by

$$x - \frac{x^2}{2} + \frac{x^3}{3} - \frac{x^4}{4}$$

find by the Chebyshev economization method an approximation of lower degree which differs from this by at most $0{\cdot}12$.

5 Given the Laguerre polynomials in Table 7.1, find the Laguerre polynomials of order 3, 4, and 5. Verify the orthogonality property for the polynomials of order 3 and 4.

6 Prove the following relations for the integrals of the Chebyshev polynomials.

$$\int T_0(x)\,dx = T_1(x), \qquad \int T_1(x)\,dx = \tfrac{1}{4}T_2(x)$$

$$\int T_r(x)\,dx = \frac{1}{2}\left[\frac{T_{r+1}(x)}{r+1} - \frac{T_{r-1}(x)}{r-1}\right]$$

8

Numerical integration

8.1 Introduction

This section is concerned with the numerical methods available for evaluating a definite integral, i.e.,

$$I = \int_a^b f(x)\, dx \tag{8.1}$$

There are various reasons why it may be necessary, or desirable, to perform this calculation by numerical approximation rather than mathematical analysis. For example, it may be either difficult or impossible to find a mathematical formula for the integral or, if the problem can be solved analytically, the function concerned may be too complicated for efficient computation. Also, an integration program for a computer library may be required which could be used for a general function without special mathematical analysis on each occasion.

   The methods which will be considered here are developed by taking a simple function $Q_n(x)$ which has the same value as $f(x)$ at a number of chosen points, x_i $(i=0,1,\ldots,n)$ and using the integral of $Q_n(x)$ as an approximation to the integral of $f(x)$. The functions $Q_n(x)$ must therefore be easy to integrate and the methods considered in this section will be based on approximation by polynomial functions which have this property. The points x_i at which we let $Q_n(x_i) = f(x_i)$ are known as the nodes of the integration formula; the values $f(x_i)$ are given the notation f_i.

   If the nodes are chosen to be equidistant then a series of formulae can be derived which are known as the Newton–Cotes formulae. The simplest two formulae of this class are the trapezoidal rule and Simpson's rule, and methods based on these formulae are often used owing to their simplicity. However, the restriction of the formulae to equidistant points loses some of the accuracy which can be obtained. Formulae which are twice as accurate as the Newton–Cotes formulae can be obtained if the nodes are specially chosen to give the maximum possible accuracy. These formulae are known as gaussian quadrature* formulae and the nodes of these formulae are the zeros of certain orthogonal polynomials. It should be noted that the formulae are only more accurate in terms of the particular mathematical definition of accuracy which is discussed below. It could easily happen that a Newton–Cotes-type formula gives a closer numerical value than a theoretically more accurate gaussian formula.

* Quadrature is another name for the calculation of definite integrals.

Another method discussed below is based on using the trapezoidal rule with several different sizes of interval and then using the technique of Richardson extrapolation to successively reduce the error. This method, known as Romberg integration, is very suitable for use in a computer program and can be shown to converge for any continuous function $f(x)$. (See Isaacson and Keller (1966) page 341.)

8.2 Newton–Cotes formulae

8.2.1 Introduction

These formulae approximate the integral

$$I = \int_a^b f(x)\,dx$$

by integrating a polynomial $Q_n(x)$ of degree n whose coefficients are chosen so that

$$Q_n(x_i) = f(x_i), \qquad i = 0, 1, \ldots, n \tag{8.2}$$

where the x_i are equidistant points. For the purposes of illustration the 'closed' formulae will be considered in which case

$$x_i = a + i.h, \qquad i = 0, 1, \ldots, n \quad \text{and} \quad h = (b-a)/n \tag{8.3}$$

The term 'closed' implies that both the end points are nodes of the integration formula. Some simple examples will indicate the motivation behind the use of

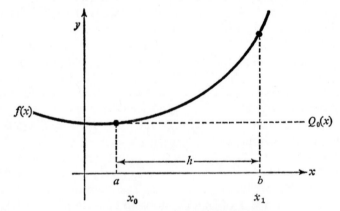

Fig. 8.1 The rectangle rule.

these formulae. The rectangle rule is based on approximation by a polynomial of degree 0, i.e., $Q_0(x) = c_0$. The area of the rectangle gives an approximation to the integral.

$$\int_a^b f(x)\,dx \approx \int_a^b Q_0(x)\,dx = hf_0 \tag{8.4}$$

This is clearly a very crude approximation and an obvious improvement would be to use a straight line between the two end points of the curve as in Fig. 8.2.

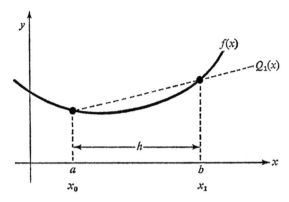

Fig. 8.2 The trapezoidal rule.

This is based on approximation by a polynomial of degree 1, i.e., $Q_1(x) = c_1 x + c_0$. The integral is found from the area of the trapezium.

$$I \approx \frac{h}{2}[f_0 + f_1] \tag{8.5}$$

Although we might expect the above method to give better results than the first one it seems reasonable that a curve would give still closer results. In this case $Q_2(x) = c_2 x^2 + c_1 x + c_0$ is used for the approximating function. The area

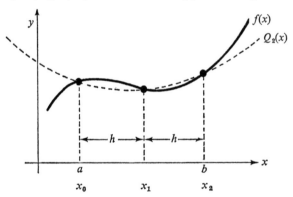

Fig. 8.3 Simpson's rule.

is found by integrating $Q_2(x)$ and substituting the values of f_i for the formula for $Q_2(x_i)$. The arithmetic detail will not be shown here but the result is

$$I \approx \frac{h}{3}[f_0 + 4f_1 + f_2] \tag{8.6}$$

A numerical example using Simpson's rule is given in Example 8.1.

Thus, we can see that by taking a series of increasing order polynomials a family of integration formulae will be generated which have the general form

$$\int_a^b f(x)\,dx = \sum_{i=0}^n a_i f_i + E \tag{8.7}$$

where we desire the error term E to be as small as possible.

8.2.2. Determination of the coefficients

We now have a general form for the approximation to the integral which has $2n + 2$ variable coefficients, i.e., the a_i and x_i, which must be chosen to make the formula as accurate as possible. In the case of the Newton–Cotes formulae some degree of choice has been sacrificed by making the x_i equally spaced, so that there are now only $n + 1$ parameters a_i to choose. It is possible to derive these coefficients from finite-difference results but it is more instructive to consider this as a general formula with various coefficients which must be chosen. The coefficients are then fixed by making the formula satisfy certain conditions leading to desirable mathematical or numerical properties. The criterion of accuracy which is used here is to make the error term zero when $f(x)$ is any polynomial of degree less than or equal to some specified value n. Since there are $n + 1$ coefficients a_i which we wish to find there are $n + 1$ conditions required to fix the coefficients. These conditions are provided by making the formula exact for the polynomials $1, x, x^2, \ldots, x^n$. It should be noted that if the formula is exact for these polynomials then it is also exact for any linear combination of these, i.e., any polynomial of degree less than or equal to n.

As an example, to show how the coefficients can be found we consider the closed Newton–Cotes formula of order 3 over the interval 0 to $3h$.

$$\int_0^{3h} f(x)\, dx = \sum_{i=0}^{3} a_i f_i + E, \qquad x_i = 0 + i.h \tag{8.8}$$

Since the formula is exact for 1, x, x^2, and x^3 we substitute these functions, in turn, into Eqn (8.8) giving four equations in four unknowns which can easily be solved.

$$f(x) = 1 \qquad 3h = a_0 + a_1 + a_2 + a_3 \tag{8.9a}$$

$$f(x) = x \qquad \frac{9h^2}{2} = h.a_1 + 2h.a_2 + 3h.a_3 \tag{8.9b}$$

$$f(x) = x^2 \qquad \frac{27h^3}{3} = h^2.a_1 + 4h^2.a_2 + 9h^2.a_3 \tag{8.9c}$$

$$f(x) = x^3 \qquad \frac{81h^4}{4} = h^3.a_1 + 8h^3.a_2 + 27h^3.a_3 \tag{8.9d}$$

These equations have the solutions

$$a_0 = \frac{3h}{8}, \qquad a_1 = \frac{9h}{8}, \qquad a_2 = \frac{9h}{8}, \qquad a_3 = \frac{3h}{8} \tag{8.10}$$

8.2.3 Discussion of errors

Intuitively, one might feel that the accuracy of these integration formulae can be increased by taking higher-order formulae until sufficient accuracy is obtained. However, there are two reasons why this strategy might fail; namely, that the function may not be adequately approximated by a polynomial, in which case the truncation error becomes large, or, that the formulae may be subject to excessive rounding error. We will now examine the formulae and their error terms and it will be seen that both these problems can arise. A detailed mathematical treatment is given in Ralston (1965) in which he shows that the general form for the Newton–Cotes integration formulae including the error term is

$$\int_a^b f(x)\,dx = A_0.h.\sum_{i=0}^{n} w_i f_i + A_1 h^{k+1} f^{(k)}(\zeta) \tag{8.11}$$

where n is the number of strips, $h = (b - a)/n$, and ζ is some value in the interval $[a, b]$. The following table gives some of the coefficients for the closed formulae. Note, that since the coefficients are symmetrical only half the table need be shown.

Table 8.1

Coefficients of the Newton-Cotes closed formulae

n	A_0	w_0	w_1	w_2	w_3	w_4	A_1
1	$\frac{1}{2}$	1	1				$-\frac{1}{12}$
2	$\frac{1}{3}$	1	4	1			$-\frac{1}{90}$
3	$\frac{3}{8}$	1	3	3	1		$-\frac{3}{80}$
4	$\frac{2}{45}$	7	32	12	32	7	$-\frac{8}{945}$
5	$\frac{5}{288}$	19	75	50	50	75	$-\frac{275}{12096}$
6	$\frac{1}{140}$	41	216	27	272	27	$-\frac{9}{1400}$
7	$\frac{7}{17280}$	751	3577	1323	2989	2989	$-\frac{8183}{518400}$
8	$\frac{4}{14175}$	989	5888	-928	10946	-4540	$-\frac{2368}{467775}$

One interesting feature which emerges from the analysis is that formulae with an even number of strips not only give zero error for polynomials up to degree n but also for polynomials of degree $n+1$. This means that in Eqn (8.11) $k = n+2$ when n is even or $k = n+1$ when n is odd. In view of this extra accuracy an even-order formula would normally be used. The exception to this is the trapezoidal rule which is valuable because of its simplicity. The other point of significance is that the error depends on a derivative of the function to be integrated. There are many simple functions whose high-order derivatives can become very large and in these cases the use of high-order formulae gives a large truncation error.

In addition to this problem of truncation error it can be seen from the table of coefficients that problems can be expected from the growth of round-off error. If we are concerned with calculating a maximum bound on the round-off error then the size of the quantity below is important.

$$B_n = A_0.h\sum_{i=0}^{n}|w_i| \tag{8.12}$$

Now, from Eqn (8.9a) we see that

$$A_0.h\sum_{i=0}^{n} w_i = b - a$$

so that in the case where all the coefficients are positive $B_n = (b - a)$. However, for the formulae where $n = 8$ and $n \geq 10$ some of the w_i are negative so that B_n is greater than $b - a$ and increases with n. Also important is the quantity

V_n which is used to estimate statistically the likely size of the error, i.e., for formulae of different order with the same interval h

$$V_n = A_0^2 \sum_{i=0}^{n} w_i^2 \tag{8.13}$$

Owing to the changes of sign in the coefficients and the large variations in size which develop in the higher-order formulae this quantity also increases with n.

8.2.4 Composite formulae

In view of the problems which may arise with high-order formulae methods must be found to reduce the error to an acceptable level. One obvious approach is to divide the interval $[a, b]$ into several sub-intervals in each of which a low-order formula is used. This reduces the size of h and, although several error terms must then be summed, the net effect is to reduce the truncation error as demonstrated below. Since the error term now depends on a low-order derivative the problems associated with large high-order derivatives have also been overcome. The coefficients of composite formulae of low order are all positive and do not vary excessively so that the quantities B_n and V_n are favourable. A discussion of the round-off error in the composite trapezoidal rule is presented in some detail in McCracken and Dorn (1964).

The formulae most often used are the composite trapezoidal rule with $h = (b - a)/m$.

$$I_m = \frac{h}{2}[f_0 + f_1 + f_1 + f_2 + \cdots + f_m]$$

$$= \frac{h}{2}[f_0 + 2f_1 + 2f_2 + \cdots + f_m] \tag{8.14}$$

and the composite Simpson's rule with $h = (b - a)/m$ where m is even.

$$I_m = \frac{h}{3}[f_0 + 4f_1 + f_2 + f_2 + 4f_3 + f_4 + \cdots + f_m]$$

$$= \frac{h}{3}[f_0 + 4f_1 + 2f_2 + 4f_3 + \cdots + f_m] \tag{8.15}$$

If we let

$$M_2 = \max_{a \leqslant x \leqslant b} f''(x) \quad \text{and} \quad M_4 = \max_{a \leqslant x \leqslant b} f^{IV}(x)$$

then the error of the composite trapezoidal rule is given by

$$|E| \leqslant \frac{mh^3}{12} M_2 = \frac{(b-a)^3}{12m^2} M_2 \tag{8.16}$$

and the error of the composite Simpson's rule is given by

$$|E| \leqslant \frac{mh^5}{180} M_4 = \frac{(b-a)^5}{180m^4} M_4 \tag{8.17}$$

Thus, for convergence of these composite formulae as m increases we require the boundedness of the second or fourth derivative respectively which can often be established. A comparison of the composite trapezoidal rule with high-order Newton–Cotes formulae is given in Example 8.5.

8.2.5 Newton–Cotes open formulae

Attention has been concentrated on the closed type of Newton–Cotes formulae since, in general, these are computationally more efficient, particularly in the composite form. However, it is sometimes convenient to use open formulae which do not use the end points a and b so that $x_i = a + i.h$ $(i = 1, \ldots, n-1$ and $h = (b-a)/n)$. These formulae find applications in the integration of ordinary differential equations as predictor formulae. The following table gives the coefficients for some low-order open formulae. The error term has the form

$$A_1 h^{k+1} f^{(k)}(\zeta) \tag{8.18}$$

where $k = n$ when n is even and $k = n-1$ when n is odd.

Table 8.2

Coefficients of the Newton–Cotes open formulae

n	A_0	w_1	w_2	w_3	w_4	A_1
2	2	1				$\frac{1}{3}$
3	$\frac{3}{2}$	1	1			$\frac{3}{4}$
4	$\frac{4}{3}$	2	-1	2		$\frac{14}{45}$
5	$\frac{5}{24}$	11	1	1	11	$\frac{95}{144}$

8.3 Gaussian quadrature

8.3.1 Introduction

If we consider the formula

$$\int_a^b f(x)\,dx = \sum_{i=0}^{n} a_i f(x_i) + E \tag{8.19}$$

we see that there are $2n + 2$ parameters, a_i and x_i, which can be varied. Since a polynomial of degree $2n + 1$ requires $2n + 2$ conditions to fix the coefficients it seems plausible that the $2n + 2$ parameters can be so chosen that all polynomials of degree less than or equal to $2n + 1$ can be integrated with zero error. We show below that this is indeed the case and investigate the method of finding the values of a_i and x_i.

If we consider the method used previously to find the coefficients a_i it can be seen that this is not easily extended to find the nodes x_i. Consider the case where $n = 1$ which leads to the following equations using the method of Section 8.2.2. If the integration limits are $a = 0$, $b = h$:

$$f(x) = 1 \qquad h = a_0 + a_1$$

$$f(x) = x \qquad \frac{h^2}{2} = a_0 x_0 + a_1 x_1$$

$$f(x) = x^2 \qquad \frac{h^3}{3} = a_0 x_0^2 + a_1 x_1^2$$

$$f(x) = x^3 \qquad \frac{h^4}{4} = a_0 x_0^3 + a_1 x_1^3 \tag{8.20}$$

Although these equations are linear in the a_i, they are non-linear in the x_i and are not easily solved. The situation would clearly become much worse for higher values of n. Fortunately, the mathematical theory associated with orthogonal polynomials provides a means of finding the nodes and the coefficients a_i can then be found by the previous method. A brief discussion of the background to gaussian quadrature is given here and further details are available in Ralston (1965) and Isaacson and Keller (1966).

8.3.2 Orthogonal polynomials

Some of the properties of orthogonal polynomials have been discussed in Chapter 7, and in the following discussion it is the fundamental property of orthogonal polynomials which is important, i.e., if there are a set of polynomials $Q_k(x)$ $(k = 0, 1, \ldots, n+1)$ which are orthogonal over the range $[a, b]$ then

$$\int_a^b Q_{n+1}(x)S_k(x)\,dx = 0, \qquad k = 0, 1, 2, \ldots, n \tag{8.21}$$

where $S_k(x)$ is any polynomial of degree k. This property holds for arbitrary $S_k(x)$ because any polynomial can be expressed as a linear combination of the orthogonal polynomials $Q_k(x)$.

We now take an arbitrary polynomial $P_{2n+1}(x)$ of degree $2n+1$ and divide by $Q_{n+1}(x)$ to give

$$P_{2n+1}(x) = Q_{n+1}(x)l_n(x) + R(x) \tag{8.22}$$

where $R(x)$ has maximum degree n. The integral can now be divided into two parts, i.e.,

$$\int_a^b P_{2n+1}(x)\,dx = \int_a^b Q_{n+1}(x)l_n(x)\,dx + \int_a^b R(x)\,dx \tag{8.23}$$

and by the orthogonality property the first integral is zero. Hence,

$$\int_a^b P_{2n+1}(x)\,dx = \int_a^b R(x)\,dx \approx \sum_{i=0}^n a_i R(x_i)$$

If we now choose the points x_i $(i = 0, 1, \ldots, n)$ as the zeros of $Q_{n+1}(x)$ then it follows that

$$P_{2n+1}(x_i) = R(x_i), \qquad i = 0, 1, \ldots, n$$

since the first term in Eqn (8.22) becomes zero.

We then have

$$\int_a^b P_{2n+1}(x)\,dx = \sum_{i=0}^n a_i P_{2n+1}(x_i) \tag{8.24}$$

where the nodes x_i are now known as the zeros of $Q_{n+1}(x)$ and the coefficients a_i can be found as in Section 8.2.2.

8.3.3 Weighted gaussian quadrature

If we use the above formulae over the range $[-1, +1]$, then the relevant orthogonal polynomials are the Legendre polynomials, but it is possible to use other orthogonal polynomials by a slight variation in the conditions of the problem. Firstly, we note that an integral with respect to x between finite limits a and b can be transformed to an integral with limits $[-1, +1]$ by the transformation

$$x = \frac{(b-a)X + a + b}{2} \tag{8.25}$$

so that the Legendre polynomials are suitable for any finite range with appropriate change of variables. Numerical results for a simple example using Legendre polynomials are given in Example 8.2. If we introduce a weighting function $w(x) > 0$ into the integrand, and use polynomials which are orthogonal with respect to this weighting function, then several different orthogonal polynomials can be used.

Let

$$\int_{-1}^{+1} f(x)\, dx = \int_{-1}^{+1} w(x) \cdot \frac{f(x)}{w(x)}\, dx$$

Hence, if $g(x) = f(x)/w(x)$ we have

$$\int_{-1}^{+1} w(x) g(x)\, dx = \sum_{i=0}^{n} a_i g(x_i) \tag{8.26}$$

As one example let us consider the orthogonal polynomials which arise when $w(x) = (1 - x^2)^{-1/2}$. These polynomials are the Chebyshev polynomials which were discussed in Chapter 7 and they have two properties which are of particular interest here. Firstly, the coefficients a_i in Eqn (8.26) all have the same value $\pi/(n+1)$ which slightly reduces the computation required and cuts down the round-off error. Also, the nodes of the Chebyshev integration formulae are given by the simple formula

$$x_j = \cos\left[\frac{(2j+1)\pi}{2(n+1)}\right], \qquad j = 0, 1, \ldots, n \tag{8.27}$$

It can be seen that if a value of the integral has been calculated for one value of n and it is necessary to increase the value of n to obtain greater accuracy it is possible to make some values of the nodes in the two formulae coincide and this will reduce the extra computation. For example, if the nodes $\cos(\pi/6)$, $\cos(3\pi/6)$, and $\cos(5\pi/6)$ were used, the degree of the formula could be multiplied by 3 to give the points $\cos(\theta_i)$ where the θ_i are given by $\pi/18$, $3\pi/18$, and $5\pi/18$. One third of these values coincide with the previous set and the $f(x_i)$ need not be recalculated. This property is not possessed by the other orthogonal polynomials.

8.3.4 Singular and infinite integrals

There are other reasons why the use of a weighting function can be advantageous. The problems considered so far have been for integrals over a finite range where values of $f(x_i)$ can be found wherever they are required. In the case where $f(x_i)$ becomes singular, even if the integral exists, the above formulae cannot be used. There are also difficulties when the derivatives of $f(x)$ become singular since this could result in an unbounded error term.

A simple example will show how the use of the weighting function can

F

avoid this difficulty. Consider the integration of the function $(1 - x^2)^{-1/4}$ over the range $[-1, +1]$. Since this function is infinite at both ends of the range it would not be possible to use a closed quadrature formula. However, using the Gauss–Chebyshev quadrature with weighting function $(1 - x^2)^{-1/2}$ we have, using Eqn (8.26)

$$\int_{-1}^{+1} \frac{(1-x^2)^{1/4}}{(1-x^2)^{1/2}} dx = \frac{\pi}{n+1} \sum_{i=0}^{n} (1 - x_i^2)^{1/4} \tag{8.28}$$

and the right-hand side function is simple to evaluate.

The use of a weighting function also allows the integration of functions over a semi-infinite or infinite range by the choice of suitable orthogonal polynomials. Over the range $[0, \infty]$ the Laguerre polynomials can be used, since they are orthogonal with respect to the weight function e^{-x} over this range. (See Example 8.3.) Over the range $[-\infty, +\infty]$ the Hermite polynomials can be used with weight function e^{-x^2}. The coefficients and nodes for some of the Laguerre and Hermite polynomials are tabulated in Ralston (1965) where further discussion on formulae with weighting functions is given.

8.4 Romberg integration

It has already been observed that the composite trapezoidal rule is simple to use and we now describe a modification to this simple rule which can lead to high accuracy and which is very convenient for computer use. For a large class of functions the trapezoidal formula for integration and its error term can be written as follows.

$$\frac{h}{2}(f_0 + 2f_1 + \cdots + f_m) = I + \sum_{j=1}^{\infty} a_j h^{2j} \tag{8.29}$$

so that the possibility exists of using Richardson extrapolation to improve the accuracy of the result (see Chapter 1). Assume, for the sake of simplicity of notation, that the initial value of m is a power of 2, say 2^k, and let the approximation given by Eqn (8.29) be designated $T_{0,k}$, i.e.,

$$T_{0,k} = I + \sum_{j=1}^{\infty} a_j h^{2j} \tag{8.30}$$

The approximation to the integral is now calculated with the interval halved, giving

$$T_{0,k+1} = I + \sum_{j=1}^{\infty} a_j \left(\frac{h}{2}\right)^{2j} \tag{8.31}$$

The first term of the error series can now be eliminated by taking a suitable combination of these two equations.

$$4I - I = 4T_{0,k+1} - T_{0,k} - \sum_{i=2}^{\infty} a_j \left[\frac{4 \cdot h^{2j}}{2^{2j}} - h^{2j} \right]$$

or

$$I = \frac{4T_{0,k+1} - T_{0,k}}{3} - \sum_{j=2}^{\infty} \frac{a_j h^{2j}}{3} \left[\frac{4}{2^{2j}} - 1 \right] \tag{8.32}$$

The first term of the right-hand side of this equation will be designated $T_{1,k}$ and we can see that the leading error term is now h^4. Successive halvings of the interval will give a sequence of values $T_{0,k}$ and each successive pair can be combined to give the values $T_{1,k}$. The sequence of values $T_{1,k}$ can then be combined in a similar manner to remove the error term in h^4 by using Richardson extrapolation. By using the formula

$$T_{p,k} = \frac{1}{4^p - 1}(4^p T_{p-1,k+1} - T_{p-1,k}), \qquad \begin{array}{l} k = 0, 1, 2, \ldots, \\ p = 1, 2, \ldots, \end{array} \qquad (8.33)$$

this process can be continued to form a sequence of columns with error terms of increasing order.

$$
\begin{array}{cccc}
h^2 & h^4 & h^6 & h^8 \\
T_{0,0} & & & \\
T_{0,1} & T_{1,0} & & \\
T_{0,2} & T_{1,1} & T_{2,0} & \\
T_{0,3} & T_{1,2} & T_{2,1} & T_{3,0}
\end{array}
\qquad (8.34)
$$

Using the fact that $h = (b - a)/2^k$ we see that the error term for the approximation $T_{p,k}$ is of the order $[(b - a)/2^k]^{2p+2}$ with each column of values converging more quickly to the true value of the integral. This method is particularly suitable for computer use since successive values can be compared to check when the process has converged. The following table of results should make this clear. Further numerical results are given in Example 8.4.

Table 8.3

Romberg integration table for sec x

Number of intervals	$T_{0,k}$	$T_{1,k}$	$T_{2,k}$	$T_{3,k}$
1	0·948 059			
2	0·899 084	0·882 759		
4	0·885 886	0·881 487	0·881 402	
8	0·882 507	0·881 381	0·881 374	0·881 372

To five decimal places the answer is 0·881 37 which agrees with the true answer 0·881 374.

8.5   Comparison of methods

In the case of functions which are available in tabulated form a formula must be chosen which is based on equidistant nodes since gaussian quadrature would involve interpolation to find the function values. It also may be convenient to use equal intervals where the data are derived from experimental observations. The choice between a high-order Newton–Cotes formula and a composite low-order formula will depend on how much is known about the derivatives of the function. In the case where high-order derivatives rapidly diminish, a high-order formula will be more efficient, assuming the accuracy of the computer is such that round-off errors do not predominate. For problems where the high-order derivatives can be quite large, the low-order composite rules would be preferred; Simpson's rule would be more accurate than the trapezoidal rule unless the fourth derivative is considerably greater than the second derivative. The most difficult problem is the choice of interval to ensure sufficient accuracy. In the situation where several similar integrations are to be performed it is worthwhile calculating several results with different intervals to find the length of interval at which sufficient accuracy is obtained. If only one integration is to be performed then the extra computation required for Romberg quadrature will be acceptable, since the sequence of values will indicate when sufficient accuracy has been obtained and it can be shown that Romberg integration will converge for any continuous function.

If the function values are available at any value of the node, as is the case when function values are calculated by computer, then the extra accuracy available from gaussian quadrature will be useful. These formulae are particularly valuable when repetitive evaluations of integrals are necessary since computational efficiency is then very important. In such a case preliminary calculations must be made to choose the order of the formula but the time taken for this is acceptable. For single calculations it is difficult to ascertain the accuracy achieved unless more than one value of the integral is obtained for checking purposes. If it was then necessary to use a higher-order formula all the nodes would in general be new values and all function values would need to be recalculated, which is not the case with equal-interval formulae. Thus gaussian quadrature formulae are not well suited to evaluations of a single integral.

Bibliographic notes

A simple introduction with numerical results for some selected low-order methods and a discussion on errors is given in McCracken and Dorn (1964). The book by Conte (1965) covers similar ground from a more mathematical viewpoint. FORTRAN programs and computer results are included. A much more detailed discussion of the methods is provided by Ralston (1965) who also provides tables of coefficients for several methods. Although this text is more mathematically demanding than those above it will be useful for some students towards the end of their undergraduate course. For those students who wish to pursue the subject in some depth the books by Davis and Rabinowitz (1967) and Krylov (1962) should be consulted. The former text includes a suite of programs in FORTRAN and a bibliography of tables of coefficients of various formulae.

A specialist text for those interested in gaussian quadrature is the book by Stroud and Secrest (1966).

Worked examples

1 In this example Simpson's rule will be used to find

$$\int_0^{\pi/4} \sec(x)\,dx$$

The correct value of the integral to six decimal places is 0·881 374.

The integral is evaluated with two, four, and eight intervals. The values of the function $\sec x$ at the various points are

x	0°	5·625°	11·25°	16·875°	22·5°
$\sec x$	1·000 000	1·004 838	1·019 591	1·044 997	1·082 392

x	28·125°	33·75°	39·375°	45°
$\sec x$	1·133 888	2·202 690	1·293 643	1·414 214

The values obtained using the composite Simpson's rule are

Number of intervals	Integral	Error	Richardson value
2	0·882 761	0·001 387	
4	0·881 489	0·000 115	0·881 404
8	0·881 383	0·000 009	0·881 376

The error term for Simpson's rule is

$$\frac{-h^4(b-a)}{180} \cdot f^{\mathrm{V}}(\zeta_1)$$

and this can be used in an attempt to improve the Simpson values by the method of Richardson extrapolation. When the step size is halved the error becomes

$$\frac{-h^4(b-a)}{16.180} \cdot f^{\mathrm{V}}(\zeta_2)$$

A new approximation is found by taking $(16I_{2n} - I_n)/15$ and the two values obtained by combining the two pairs of results are shown. The error in each case is reduced to about one quarter.

The ideal way of choosing a step size is to calculate the maximum value of the fifth-order derivative in the error formula and use this to calculate a suitable value of h to make the sum of the error terms less than some chosen value. In many examples, as in this case, the fifth derivative has a complicated formula and finding the bound is not a trivial matter.

2 The Gauss–Legendre integration formula with three nodes will be derived and it will be shown that this formula is accurate for a polynomial of degree 5. The Legendre polynomial of degree 3 is $\frac{1}{2}(5x^3 - 3x)$ which has roots

$$0, \pm\sqrt{(0\cdot6)} = 0, \pm 0\cdot774\,597$$

If we substitute the polynomials $1, x, \ldots$, we obtain equations for the coefficients in the gaussian formula

$$\int_{-1}^{+1} f(x)\,dx = \sum_{i=0}^{2} a_i f(x_i)$$

Since, $x_0 = -0\cdot774\,597$, $x_1 = 0$, and $x_2 = 0\cdot775\,497$

$$
\begin{aligned}
f(x) &= 1 & 2 &= a_0 + a_1 + a_2 \\
f(x) &= x & 0 &= a_0 x_0 + a_2 x_2 & a_0 &= a_2 \\
f(x) &= x^2 & \tfrac{2}{3} &= a_0 x_0^2 + a_2 x_2^2 & a_0 + a_2 &= \tfrac{10}{9} \\
f(x) &= x^3 & 0 &= a_0 x_0^3 + a_2 x_2^3 & a_0 &= a_2 = \tfrac{5}{9} \\
f(x) &= x^4 & \tfrac{2}{5} &= a_0 x_0^4 + a_2 x_2^4 \\
f(x) &= x^5 & 0 &= a_0 x_0^5 + a_2 x_2^5 \\
\end{aligned}
$$

$$a_0 = a_2 = \tfrac{5}{9}, \qquad a_1 = \tfrac{8}{9}$$

We see that

$$\int_{-1}^{+1} x^5\,dx = 0 = -\tfrac{5}{9}(0\cdot774\,597)^5 + \tfrac{5}{9}(0\cdot774\,597)^5$$
$$= 0$$

3 The Gauss–Laguerre formulae are used for integrals of the form

$$\int_0^\infty e^{-x}f(x)\,dx = \sum_{i=0}^n a_i f(x_i)$$

The coefficients and nodes for the formula of order 4 are given below.

Nodes	Coefficients	Product
0·322 548	0·603 154	0·194 546
1·745 761	0·357 419	0·623 968
4·536 620	0·038 888	0·176 420
9·395 071	0·000 539	0·005 063
	1·000 000	0·999 997

If $f(x) = 1$ the value of the integral is 1 and the value of the approximation

is

$$0·603\,154 + 0·357\,419 + 0·038\,888 + 0·000\,539 = 1·000\,000$$

which is exact as the theory predicts. If $f(x) = x$ the value of the integral is

$$\left[-e^{-x}x\right]_0^\infty + \left[\frac{e^{-x}}{-1}\right]_0^\infty = 1·0$$

The value of the approximation is 0·999 998 which is exact within the limits imposed by using only six decimal places. Consider an integral which does not involve a polynomial $f(x)$.

$$\int_0^\infty e^{-x}\sin x\,dx = \sum_{i=0}^3 a_i \sin(x_i)$$

The true value of the integral is 0·5. The approximation gives the value

$$0·603\,154 \times 0·316\,984 + 0·357\,419 \times 0·984\,732 + 0·038\,888 \times (-0·984\,593)$$
$$+ 0·000\,539 \times 0·029\,703 = 0·504\,879$$

Thus, we have an error of approximately 1 % for a formula of order as low as 4. This is remarkable in view of the fact that $\sin x$ is poorly approximated by a polynomial over the range in question.

4 The following Romberg integration table shows the convergence of the method. The method is clearly suitable for this problem since the true solution is $1·000\,000 = \int_0^{\pi/2} \sin x\,dx$

Number of intervals	$T_{0,k}$	$T_{1,k}$	$T_{2,k}$	$T_{3,k}$
1	0·785 398			
2	0·948 059	1·002 279		
4	0·987 116	1·000 135	0·999 992	
8	0·996 785	1·000 008	1·000 000	1·000 000
16	0·999 197	1·000 001	1·000 000	1·000 000

The use of Aitken's Δ^2 process is also shown. If the last three values of the table are used we have

$f(x)$	$\Delta f(x)$	$\Delta^2 f(x)$
0·987 116		
0·991 762	0·004 646	
0·994 282	0·002 520	−0·002 126

Hence, the improved value is found from

$$f = f_n - \frac{(\Delta f_n)^2}{\Delta^2 f_n} = 0.987\,116 - \frac{(0.004\,646)^2}{-0.002\,126}$$
$$= 0.997\,271$$

Problems

1 The following formula is to be used for the numerical approximation of an integral.

$$\int_a^b f(x)\,dx \approx \sum_{i=0}^n a_i f(x_i)$$

The criterion for accuracy of this formula is that the formula shall be exact for all polynomials of degree less than or equal to n. If $a = 0$ and $b = 6$ and the values of $f(x)$ are available at the points $x_0 = 0$, $x_1 = 1$, $x_2 = 5$, and $x_3 = 6$, find the values of the coefficients a_i which satisfy the above requirement of accuracy.

2 A formula with the general form

$$I = a_0 f(x_0) + a_1 f(x_1) + a_2 f(x_2)$$

is to be used to give an approximation for the integral of $f(x)$ between x_0 and x_2 where $x_n = x_0 + n.h$. The criterion for accuracy is that the formula shall be exact for polynomials of as high a degree as possible. Find the coefficients a_0, a_1, and a_2. Is the formula exact when integrating a cubic?

3 Find the polynomial of degree 2 which is orthogonal to the polynomial 1 and x over the range $[-2, 2]$. Hence, find the nodes x_i and the weights a_i in the gaussian integration formula

$$I = \sum_{i=0}^1 a_i f(x_i)$$

4 Check the above results by using the criterion that the gaussian quadrature shown is accurate for all polynomials up to degree 3 as a basis for deriving the nodes and weights in the formula.

5 Evaluate

$$\int_0^{0.8} \sin x\,dx$$

using the composite trapezoidal rule at intervals $h = 0.4$.

Compare this with the result using Simpson's rule with the same interval. Is the trapezium rule, with twice the number of intervals, more accurate than Simpson's rule? How effective is the use of Richardson's extrapolation on the values for the integral obtained from the above two applications of the trapezium rule.

9

Eigenvalues and eigenvectors

9.1 The fundamental equations

There are many physical problems which require for their solution the values of λ which satisfy the equation

$$AX = \lambda X \tag{9.1}$$

where $\mathbf{A}$ is an $n \times n$ matrix, λ is a scalar known as the eigenvalue, and $\mathbf{X}$ is the corresponding eigenvector. Such problems occur in the analysis of stability and in the equations of vibrations in structures or electrical circuits. An example is provided by the vibrations of a body disturbed slightly from an equilibrium position. The equations have the following typical form.

$$a_{11}\ddot{x} + b_{11}x + a_{12}\ddot{y} + b_{12}y = 0$$
$$a_{21}\ddot{x} + b_{21}x + a_{22}\ddot{y} + b_{22}y = 0 \tag{9.2}$$

The solutions are provided by $x = K_1 \cos(pt + \varepsilon_1)$ and $y = K_2 \cos(pt + \varepsilon_2)$, and substitution of these values in the equations gives

$$-p^2 \begin{bmatrix} a_{11} & a_{12} \\ a_{21} & a_{22} \end{bmatrix} \begin{bmatrix} x \\ y \end{bmatrix} + \begin{bmatrix} b_{11} & b_{12} \\ b_{21} & b_{22} \end{bmatrix} \begin{bmatrix} x \\ y \end{bmatrix} = 0 \tag{9.3}$$

Putting the equations in matrix form, we require the values of $\lambda \equiv p^2$ which satisfy the equation

$$\lambda AX = BX$$

or

$$A^{-1}BX = \lambda X \tag{9.4}$$

Since Eqn (9.1) can be rewritten in the form $(\mathbf{A} - \lambda \mathbf{I})\mathbf{X} = \mathbf{0}$ we see that Eqn (9.1) has a solution where the components of $\mathbf{X}$ are nonzero if, and only if, the determinant of $\mathbf{A} - \lambda \mathbf{I}$ is zero, i.e.,

$$\begin{vmatrix} a_{11} - \lambda & a_{12} & \cdots & a_{1n} \\ a_{21} & a_{22} - \lambda & \cdots & a_{2n} \\ \cdots & \cdots & \cdots & \cdots \\ a_{n1} & \cdots & \cdots & a_{nn} - \lambda \end{vmatrix} = 0 \tag{9.5}$$

It can be seen that the expansion of this determinant would give a polynomial in λ of degree n and the roots of this equation would give the required eigenvalues. This polynomial equation is known as the characteristic equation of the matrix

A. The solution of the homogeneous equation (9.1) then gives the eigenvector **X**. In the case where the eigenvalues are all distinct there is a unique eigenvector corresponding to each eigenvalue. In the case of an eigenvalue of multiplicity m there may be m corresponding eigenvectors or less. In the latter case there will be less than n eigenvectors for the matrix **A** and the eigenvectors cannot form a base for the space. This would cause difficulties when trying to use the iterative methods of Section 9.3.

The discussion of Chapter 3 on the solution of polynomial equations emphasized the difficulty of obtaining satisfactory solutions to this problem. In the present problem the coefficients of the polynomial would be found by some process of elimination, or evaluation of determinants, and this could mean the coefficients of the characteristic polynomial contain significant errors. In view of these difficulties the method of expansion as a polynomial, followed by a subsequent solution, is not satisfactory except for very low-order matrices.

There are two types of method which can be used. The iterative methods are very simple to use and, in certain circumstances, one root can be found very effectively. These methods can be modified to find more than one root but in general they would not be used to find all the eigenvalues of a matrix. The methods which are used when all the eigenvalues of a matrix are required are based on transformations which reduce the matrix to a simpler form which can easily be solved to find the eigenvalues. If similarity transformations are used the new matrices have the same eigenvalues as the original matrices and a simple relationship exists between the old and the new eigenvectors.

9.2 Some useful matrix results

It will be useful to present some parts of matrix theory which form the background to the work of this chapter. It is, however, assumed that the reader has some acquaintance with matrix calculus including eigenvalue and eigenvector theory. In this chapter it is assumed that the matrix **A** has real coefficients. The following results are of interest.

1 A matrix is said to be symmetric if $a_{ij} = a_{ji}$ $(i,j = 1, 2, \ldots, n)$. The following examples illustrate this point.

$$\begin{bmatrix} 1 & -2 & 1 & -3 \\ -2 & 2 & -3 & 0 \\ 1 & -3 & -4 & 1 \\ -3 & 0 & 1 & 2 \end{bmatrix}, \qquad \begin{bmatrix} 3 & -5 & 0 & 1 \\ 3 & 1 & -3 & 0 \\ -2 & 2 & 1 & -5 \\ -1 & -2 & 3 & 3 \end{bmatrix}$$

Symmetric   *Not symmetric*

2 A matrix is said to be orthogonal if

$$\mathbf{a}_i^T \mathbf{a}_j = 0, \qquad i \neq j$$
$$\mathbf{a}_i^T \mathbf{a}_i = 1 \tag{9.6}$$

where the vector $\mathbf{a}_i$ is the ith column of **A**. It should be noted that the inverse of an orthogonal matrix can be obtained without calculation since $\mathbf{A}^T \mathbf{A} = \mathbf{I}$ according to Eqns (9.6) and hence $\mathbf{A}^{-1} = \mathbf{A}^T$

G

3 The eigenvalues of a real symmetric matrix are real. The eigenvectors corresponding to distinct eigenvalues are orthogonal.

4 If a matrix of order n has n distinct eigenvalues then there are n linearly independent eigenvectors which can form a base for the space of vectors. An arbitrary vector can then be expressed in terms of the eigenvectors, i.e.,

$$y = \sum_{r=1}^{n} a_r X^{(r)} \tag{9.7}$$

where $X^{(r)}$ $(r = 1, 2, \ldots, n)$ are the linearly independent eigenvectors.

5 If $X^{(l)}$ is an eigenvector corresponding to the eigenvalue λ_l then $AX^{(l)} = \lambda_l X^{(l)}$ and $A^k X^{(l)} = \lambda_l^k X^{(l)}$. Thus, the effect of successive multiplication of an eigenvector by the matrix A is to successively multiply the vector by the scalar λ_l.

6 Two matrices, A and B are said to be similar if a non-singular matrix P exists such that $B = P^{-1}AP$. It is easy to see that similar matrices have the same eigenvalues since if

$$AX = \lambda X$$

then

$$P^{-1}AX = \lambda P^{-1}X$$

and, if

$$X = PY \tag{9.8}$$

then

$$P^{-1}APY = \lambda Y \tag{9.9}$$

The eigenvectors of A can be found from the eigenvectors of B by the relation $X = PY$.

7 If $X^{(r)}$ is an eigenvector of a matrix then any scalar multiple of this is also an eigenvector. It will sometimes be convenient to normalize the eigenvector and this can be done in two ways. One method of normalization is to divide all the elements of a vector by the largest element so that vectors have unity as the largest element. Alternatively, each element could be divided by the sum of the squares of the elements of the vector in which case vectors have unit length.

9.3 Iterative methods

9.3.1 Finding the real eigenvalue of largest modulus

Iterative methods are most suitable for problems in which only one or two roots are to be found, although there are ways of extending these methods to find all roots. In this section it will be assumed that all the eigenvalues are distinct. Therefore, the eigenvectors are linearly independent and an arbitrary vector can be expressed in the form

$$y = \sum_{r=1}^{n} a_r X^{(r)} \tag{9.10}$$

To find the largest eigenvalue and its corresponding eigenvector an arbitrary vector $y^{(0)}$ is multiplied successively by the matrix A. In general, the sequence of vectors $y^{(k)}$ will converge to the eigenvector and the ratio of elements of successive vectors will converge to the eigenvalue. For example, let λ_1 be the largest eigenvalue and $X^{(1)}$ the corresponding eigenvector. Then,

$$y^{(k)} = A^k y^{(0)} \qquad (9.11)$$

$$= A^k \sum_{r=1}^{n} a_r X^{(r)}$$

$$= \sum_{r=1}^{n} a_r \lambda_r^k X^{(r)}$$

$$= \lambda_1^k \left[a_1 X^{(1)} + a_2 \left(\frac{\lambda_2}{\lambda_1} \right)^k X^{(2)} + \cdots + \left(\frac{\lambda_n}{\lambda_1} \right)^k X^{(n)} \right] \qquad (9.12)$$

The values of $(\lambda_r/\lambda_1)^k$ $(r \neq 1)$ tend to zero as k tends to ∞ and so all the terms become negligible except the first term. Thus, $y^{(k)}$ tends to a scalar multiple of $X^{(1)}$ and the ratio of an element $y_i^{(k+1)}$ to the corresponding element $y_i^{(k)}$ tends to λ_1 as k increases.

It can be seen that convergence will be quicker if the initial vector contains a large component of $X^{(1)}$. If the component $X^{(1)}$ is missing completely, i.e., $a_1 = 0$, then it appears that the sequence cannot converge to the dominant eigenvector. However, round-off error will normally produce a component of $X^{(1)}$ which will then become magnified and eventually become dominant. If convergence appears slow, with a particular choice of initial vector, then more satisfactory progress can sometimes be made by choosing a different initial vector. The rate of convergence is clearly affected by the ratio of the moduli of the two largest eigenvalues. When this ratio is nearly unity very poor convergence will result.

The computational procedure is slightly different from the process described above so that unbounded growth of the elements of $y^{(k)}$ can be avoided.

1 The vector $y^{(0)}$ is put into normalized form with the largest element unity.

2 The vector is multiplied by the matrix A.

3 The new vector is normalized by dividing each element by the largest element which we shall designate q_k.

4 The vector $y^{(k)}$ is repeatedly multiplied by the matrix A and divided by the factor q_k until the values of q_k and q_{k+1} differ by some specified small value. The value of q_k gives the value of the largest eigenvalue and the vector $y^{(k)}$ is the eigenvector corresponding to this.

Computer results for this method are given in Example 9.1.

9.3.2 Complex-conjugate roots

If the root of largest modulus is complex then, since the matrix A is real, there will be another root of equal modulus which is the complex conjugate. The previous analysis is no longer relevant and multiplication by the matrix A will not produce a sequence of vectors which converge. If the arbitrary initial vector $y^{(0)}$ is expanded in terms of the eigenvalues as before then, after several multiplications by the matrix A, all terms except two will be negligible. Hence,

$$y^{(k)} \approx a_1 \lambda_1^k X^{(1)} + a_2 \bar{\lambda}_1^k \bar{X}^{(1)} \qquad (9.13)$$

The following analysis shows how the eigenvalues can be found. Let λ_1 and $\bar{\lambda}_1$ be the solutions of the quadratic equation

$$\lambda^2 + a\lambda + b = 0 \qquad (9.14)$$

where a and b are unknown quantities at this stage. Three consecutive vectors of the sequence are taken $\mathbf{y}^{(k)}$, $\mathbf{y}^{(k+1)}$, and $\mathbf{y}^{(k+2)}$ and the following linear combination is formed.

$$
\begin{aligned}
\mathbf{y}^{(k+2)} &+ a\mathbf{y}^{(k+1)} + b\mathbf{y}^{(k)} \\
\approx a_1\lambda_1^{k+2}\mathbf{X}^{(1)} &+ a_2\bar{\lambda}_1^{k+2}\bar{\mathbf{X}}^{(1)} + a[a_1\lambda_1^{k+1}\mathbf{X}^{(1)} + a_2\bar{\lambda}_1^{k+1}\bar{\mathbf{X}}^{(1)}] \\
&+ b[a_1\lambda_1^{k}\mathbf{X}^{(1)} + a_2\bar{\lambda}_1^{k}\bar{\mathbf{X}}^{(1)}] \\
= a_1\lambda_1^k\mathbf{X}^{(1)}[\lambda_1^2 + a\lambda_1 + b] &+ a_2\bar{\lambda}_1^k\bar{\mathbf{X}}^{(1)}[\bar{\lambda}_1^2 + a\bar{\lambda}_1 + b] = 0 \qquad (9.15)
\end{aligned}
$$

Thus, knowing a and b, a linear combination of three consecutive vectors can be formed which approximately equals the null vector. In the problem being considered we wish to perform the inverse calculation, i.e., we try to find values of a and b such that a linear combination of three consecutive vectors gives the null vector. If any two components of these vectors are taken, and the linear combination equated to zero, then two equations are generated for the two unknowns a and b and these can normally be solved

$$
\begin{aligned}
y_r^{(k+2)} + ay_r^{(k+1)} + by_r^{(k)} &= 0 \\
y_s^{(k+2)} + ay_s^{(k+1)} + by_s^{(k)} &= 0
\end{aligned}
\qquad (9.16)
$$

Once the values of a and b are known, then Eqn (9.14) can be solved to give the values of λ_1 and $\bar{\lambda}_1$. The eigenvector $\mathbf{X}^{(1)}$ is then found from two consecutive vectors

$$\mathbf{y}^{(k+1)} - \bar{\lambda}_1\mathbf{y}^{(k)} = a_1\lambda_1^k[\lambda_1 - \bar{\lambda}_1]\mathbf{X}^{(1)} \qquad (9.17)$$

and $\bar{\mathbf{X}}^{(1)}$ is found in a similar manner.

$$\mathbf{y}^{(k+1)} - \lambda_1\mathbf{y}^{(k)} = a_2\bar{\lambda}_1^k[\bar{\lambda}_1 - \lambda_1]\bar{\mathbf{X}}^{(1)} \qquad (9.18)$$

Practically, the iteration process is continued until the values of a and b become effectively constant, no matter which components of the vectors $\mathbf{y}^{(k)}$ are used, and also the values do not change from one iteration to the next. Computer results for this method are given in Example 9.3.

9.3.3 Inverse iteration for the smallest real root

By a simple modification to the above method it is possible to use iteration to find the smallest eigenvalue. The penalty is that the new process involves the repeated solution of sets of linear simultaneous equations. The process is based on the property that the eigenvalues of $\mathbf{A}^{-1}$ are the inverses of the eigenvalues of $\mathbf{A}$; therefore, the smallest eigenvalue of $\mathbf{A}$ is the largest eigenvalue of $\mathbf{A}^{-1}$. The eigenvectors of $\mathbf{A}$ and $\mathbf{A}^{-1}$ are the same.

We therefore iterate with $\mathbf{A}^{-1}$, i.e.,

$$\mathbf{y}^{(k)} = \mathbf{A}^{-1}\mathbf{y}^{(k-1)} \qquad (9.19)$$

However, in Chapter 4 it was shown that the process of finding $\mathbf{A}^{-1}$

explicitly was inefficient and should be replaced by gaussian elimination. This can be done in this problem by finding the sequence of vectors $y^{(k)}$ by solving a succession of equations.

$$\mathbf{A}y^{(k)} = y^{(k-1)}, \qquad k = 1, 2, \ldots \tag{9.20}$$

After the first reduction to triangular form, the next solutions involve multiplication by a triangular matrix and back substitution so that the computation time is reduced. After each solution the vector $y^{(k)}$ is normalized by dividing by the element of largest modulus q_k. The process is stopped when the difference between successive values of q_k is less than some specified amount. The value of the smallest eigenvalue of $\mathbf{A}$ is then given by q_k^{-1} and the eigenvector by $y^{(k)}$.

9.3.4 Finding the root nearest a given value
Let the matrix $\mathbf{B} = \mathbf{A} - p\mathbf{I}$ where the eigenvalue nearest to p is required. Eigenvectors of $\mathbf{A}$ satisfy $\mathbf{A}\mathbf{X} = \lambda\mathbf{X}$ and therefore they also, satisfy the equation

$$(\mathbf{A} - p\mathbf{I})\mathbf{X} = (\lambda - p)\mathbf{X} \tag{9.21}$$

It follows that the eigenvectors of $\mathbf{B}$ and $\mathbf{A}$ are the same and the new eigenvalues are $\lambda - p$. Hence, the eigenvalue λ_i nearest to p will correspond to the smallest eigenvalue of $\mathbf{B}$ and the method of the previous section can be used to find $(\lambda_i - p)^{-1}$. If the divisor is q_k as before then $\lambda_i = p + 1/q_k$ and the eigenvector is equal to $y^{(k)}$. Computer results for this method are given in Example 9.2.

9.3.5 Extension of the method
The iterative method is most suitable in circumstances where the largest or smallest eigenvalue, or one particular eigenvalue is required and there are several physical problems which are of this type. A process for extending the above scheme to find all the eigenvalues of a matrix will now be described. However, the process described is not recommended for large matrices since the build-up of errors can be very large. The basis of the method is to eliminate in some way the largest eigenvalue and eigenvector so that the iteration scheme of Section 9.2.1 now picks out the next largest eigenvalue. The process is known as deflation and two methods will be described, one which is applicable to any matrix, and a second which is restricted to symmetric matrices.

Let λ_1 be the largest eigenvalue and $\mathbf{X}^{(1)}$ the corresponding eigenvector. The matrix $\mathbf{A}$ is partitioned according to which element of $\mathbf{X}^{(1)}$ is the largest. Assume that the first element is the largest in which case the matrix $\mathbf{A}$ is represented by

$$\mathbf{A} = \begin{pmatrix} \mathbf{a}_1 \\ \mathbf{B} \end{pmatrix} \tag{9.22}$$

where $\mathbf{a}_1$ is the first row of $\mathbf{A}$, and $\mathbf{B}$ is the $n-1 \times n$ matrix of the remaining rows. Since the largest component of the vector $\mathbf{X}^{(1)}$ is the first, all the vectors in the discussion will be normalized so that the first component is unity.

A matrix

$$\mathbf{A}_1 = \mathbf{A} - \mathbf{X}^{(1)}\mathbf{a}_1 \tag{9.23}$$

is formed. It will be shown that A_1 has eigenvectors λ_i $(i = 2, 3, \ldots, n)$ which are the same as A and the remaining eigenvalue equals zero.

Consider any other eigenvalue λ_i of A with corresponding eigenvector $X^{(i)}$.

$$A_1(X^{(1)} - X^{(i)}) = A(X^{(1)} - X^{(i)}) - X^{(1)}a_1(X^{(1)} - X^{(i)}) \qquad (9.24)$$

Now, since a_1 is the first row of the matrix A, the product $a_1X^{(i)}$ is the first element of the vector $\lambda_i X^{(i)}$. This equals λ_i since the vectors are normalized. Hence, the right-hand side of Eqn (9.24) equals

$$\lambda_1 X^{(1)} - \lambda_i X^{(i)} - X^{(1)}(\lambda_1 - \lambda_i) \qquad (9.25)$$
$$= \lambda_i(X^{(1)} - X^{(i)})$$

This shows that the eigenvalues λ_i $(i = 2, 3, \ldots, n)$ are eigenvalues of the new matrix and the corresponding eigenvectors are $X^{(1)} - X^{(i)}$. Also,

$$A_1 X^{(1)} = AX^{(1)} - X^{(1)}a_1 X^{(1)}$$
$$= AX^{(1)} - \lambda_1 X^{(1)} = 0 \qquad (9.26)$$

so that the remaining eigenvalue is zero. The matrix A_1 can now be used to iterate for the next largest eigenvalue and so on.

It should be noted that a simplification can be employed to reduce the computation involved. All the eigenvectors of the new matrix have zero as the first component and, therefore, only $n - 1$ components of the eigenvectors need be found. Also, the matrix A_1 must have zero elements in the first row since the first rows of A and $X^{(1)}a_1$ are both a_1. Therefore, the problem can be solved as a problem of dimension $n - 1$ by striking out the first row and column of the matrix A_1.

When the eigenvector $r^{(2)}$, corresponding to the largest eigenvalue of the matrix A_1, has been found this can be used to find the eigenvector of A, corresponding to the eigenvalue λ_2. It is assumed that λ_2 is the next largest eigenvalue. From the previous discussion we have that $r^{(2)}$ is a scalar multiple of $X^{(1)} - X^{(2)}$, i.e.,

$$cr^{(2)} = X^{(1)} - X^{(2)}$$

Hence,

$$X^{(2)} = X^{(1)} - cr^{(2)} \qquad (9.27)$$

where c is a constant which must be found. Also,

$$a_1 X^{(2)} = a_1 X^{(1)} - ca_1 r^{(2)}$$
$$\lambda_2 = \lambda_1 - ca_1 r^{(2)}$$

and this gives the value for c.

$$c = (\lambda_1 - \lambda_2)/a_1 r^{(2)} \qquad (9.28)$$

which can then be inserted in Eqn (9.27) to give $X^{(2)}$.

In the symmetric case, an alternative method exists which is less susceptible to round-off errors. In this analysis the vectors are normalized by dividing by the sum of the squares of the elements of the vector. This makes the length of the vectors equal to unity. Form the matrix

$$A_1 = A - \lambda_1 X^{(1)}[X^{(1)}]^T \tag{9.29}$$

A_1 has the same eigenvalues as A, namely λ_i ($i = 2, 3, \ldots, n$) and the remaining root of A_1 is zero. This follows since

$$A_1 X^{(i)} = A X^{(i)} - \lambda_1 X^{(1)}[X^{(1)}]^T X^{(i)}$$
$$= \lambda_i X^{(i)}, \qquad i \neq 1 \tag{9.30}$$

since the eigenvectors of a symmetric matrix are orthogonal. Also,

$$A_1 X^{(1)} = A X^{(1)} - \lambda_1 X^{(1)}[X^{(1)}]^T X^{(1)}$$
$$= A X^{(1)} - \lambda_1 X^{(1)}$$
$$= 0 \tag{9.31}$$

The above analysis also shows that the eigenvectors corresponding to the eigenvalues λ_i ($i = 2, 3, \ldots, n$) are the same for the matrices A and A_1. This method is shown in the numerical example (9.1).

9.4 Transformation methods

9.4.1 Jacobi's method

The methods of this section use similarity transformations to obtain a transformed matrix with the same eigenvalues but of simpler form. The transformation matrices which are used are orthogonal matrices, since it can be shown that such matrices are the most suitable for minimizing the errors in the process.

The most simple form which could be achieved is the diagonal form since the eigenvalues would then be available directly as the diagonal elements. The Jacobi method is designed to produce a diagonal form by systematically eliminating the off-diagonal elements. It is, however, an iterative process requiring an infinite number of steps. This introduces two disadvantages, firstly, the process may converge very slowly or not at all and, secondly, the need to truncate the process may introduce errors which seriously disturb the correct solutions.

The computational scheme is straightforward. A new matrix $A_1 = P_1^{-1} A P_1$ is formed using a matrix P_1 which introduces a zero off-diagonal element in A_1. A further matrix A_2 is produced, $A_2 = P_2^{-1} A_1 P_2$, in which a new off-diagonal zero is produced. Unfortunately, in the Jacobi process, the introduction of each new zero usually introduces a new element into the previous zero position. The process is continued either by working systematically along one row and then the next, introducing zero elements, or alternatively, by eliminating the largest off-diagonal element at each stage. When all off-diagonal elements are less in modulus than some specified small quantity the process is terminated. The eigenvalues are then taken to be the diagonal elements.

Since the computation is performed with the orthogonal matrices, there is no need to calculate the inverse which is given by $P_r^{-1} = P_r^T$. The final matrix is

$$A_r = P_r^T P_{r-1}^T \cdots P_1^T A P_1 \cdots P_{r-1} P_r \tag{9.32}$$

and, if $\mathbf{Y}^{(r)}$ is the eigenvector of $\mathbf{A}_r$, the eigenvector of the original matrix $\mathbf{A}$ is

$$\mathbf{P}_1 \cdots \mathbf{P}_{r-1} \mathbf{P}_r \mathbf{Y}^{(r)} \tag{9.33}$$

The orthogonal matrices which are used in the Jacobi and Givens methods are extensions of a rotation matrix in a two-dimensional system. The $n \times n$ rotation matrix to rotate in the (r, s) plane is given by the $n \times n$ unit matrix with the following four changes.

$$a_{rr} = \cos \theta, \qquad a_{rs} = -\sin \theta$$
$$a_{sr} = \sin \theta, \qquad a_{ss} = \cos \theta$$

For example, the following matrix corresponds to a rotation in the $(2, 3)$ plane.

$$\mathbf{P}_1 = \begin{bmatrix} 1 & 0 & 0 & 0 \\ 0 & c & -s & 0 \\ 0 & s & c & 0 \\ 0 & 0 & 0 & 1 \end{bmatrix}, \qquad \begin{array}{l} c = \cos \theta \\ s = \sin \theta \end{array} \tag{9.34}$$

The transformation $\mathbf{P}_1^T \mathbf{A} \mathbf{P}_1$ gives the matrix

$$\begin{array}{cccc} a_{11} & ca_{12} + sa_{13} & -sa_{12} + ca_{13} & a_{14} \\ ca_{21} + sa_{31} & c^2a_{22} + csa_{23} + csa_{32} + s^2a_{33} & -csa_{22} + c^2a_{23} - s^2a_{32} + csa_{33} & ca_{24} + sa_{34} \\ -sa_{21} + ca_{31} & -csa_{22} - s^2a_{23} + c^2a_{32} + csa_{33} & s^2a_{22} - csa_{23} - csa_{33} + c^2a_{32} & ca_{34} - sa_{24} \\ a_{41} & ca_{42} + sa_{43} & -sa_{42} + ca_{43} & a_{44} \end{array}$$
$$\tag{9.35}$$

The Jacobi method reduces one coefficient to zero by choosing the value of θ so that the element in the position 2, 3 becomes zero. The method is normally used for symmetric matrices so that it is required that

$$(c^2 - s^2)a_{23} + cs(a_{33} - a_{22}) = 0 \quad \text{or} \quad \tan 2\theta = \frac{2a_{23}}{a_{22} - a_{33}} \tag{9.36}$$

Computer results for this method are given in Example 9.4.

9.4.2 Givens' method

Givens' method is based on matrix transformations of the same kind as in Jacobi's method, but the scheme is designed so that any zeros which are created are retained in the subsequent transformations. When a rotation is performed in the plane (r, s) the element $(r - 1, s)$ is eliminated for $r = 1, 2, \ldots, n - 1$ and $s = r + 2, r + 3, \ldots, n$. Thus, for the above rotation in the 2, 3 plane the value of θ would be chosen to satisfy.

$$-sa_{12} + ca_{13} = 0$$
$$\tan \theta = \frac{a_{13}}{a_{12}}$$

or, more generally

$$\tan \theta = \frac{a_{r-1,s}}{a_{r-1,r}} \tag{9.37}$$

It can be seen that the main diagonal elements, and the diagonal elements immediately above and below the main diagonal elements remain non-zero. Thus, the end result of the process is not the simple diagonal form but the so-called tridiagonal form.

$$
\begin{bmatrix}
x & x & & & & & \\
x & x & x & & & 0 & \\
  & x & x & x & & & \\
\multicolumn{7}{c}{\cdots\cdots\cdots\cdots\cdots\cdots\cdots\cdots} \\
  & & & x & x & x & \\
  & 0 & & & x & x & x \\
  & & & & & x & x \\
\end{bmatrix}
\tag{9.38}
$$

The eigenvalues of the tridiagonal form cannot be found immediately, as in the diagonal case. However, the method of solution is simple enough to make the reduction to tridiagonal form a worthwhile computational step. This method is described in Section 9.5.

It can easily be seen that the zeros will be conserved if the elimination is done systematically along the first row starting from element 1, 3, and then along the second row starting from element 2, 4, etc. Computer results for this method are given in Example 9.5.

If the matrix **A** is not symmetric Givens' method could still be applied. However, the final form would not be the symmetric tridiagonal form but the Hessenberg form.

$$
\begin{bmatrix}
x & x & & & \\
x & x & x & & 0 \\
\multicolumn{5}{c}{\cdots\cdots\cdots\cdots\cdots} \\
\multicolumn{5}{c}{\cdots\cdots\cdots\cdots\cdots} \\
\cdots\cdots\cdots & x & x & x \\
\cdots\cdots\cdots & & x & x \\
\end{bmatrix}
\tag{9.39}
$$

9.4.3 Householder's method

Although Givens' method was a considerable advance on the method of Jacobi, this has been superseded by a method due to Householder. This method also uses orthogonal transformations to reduce a symmetric matrix to tridiagonal form or an unsymmetric matrix to Hessenberg form. The advantage of the method is that all the zeros possible on a single row are produced by a single transformation. The Householder method therefore takes only $n-2$ similarity transformations compared with $(n^2 - 3n + 2)/2$ for Givens' method. Householder's method is more complicated computationally, but there is still a substantial saving on computer time. Also, the reduction of the number of computer operations reduces error propagation.

The method is defined by the equations

$$\begin{aligned}
\mathbf{A}_0 &= \mathbf{A} \\
\mathbf{A}_r &= \mathbf{P}_r^T \mathbf{A}_{r-1} \mathbf{P}_r, \qquad r = 1, 2, \ldots
\end{aligned}$$

(9.40)

where

$$\mathbf{P}_r = \mathbf{I} - 2\omega^{(r)}[\omega^{(r)}]^T, \qquad [\omega^{(r)}]^T\omega^{(r)} = 1$$

(9.41)

As an example, the matrix $\mathbf{P}_1$ which will introduce zeros in positions $(1, 3)$ and $(1, 4)$ uses

$$(\omega^{(1)})^T = (0, \omega_2^{(1)} \, \omega_{(3,}^{(1)} \, \omega_4^{(1)})$$

$$\mathbf{P}_1 = \begin{bmatrix}
1 & 0 & 0 & 0 \\
0 & 1 - 2[\omega_2^{(1)}]^2 & -2[\omega_2^{(1)}][\omega_3^{(1)}] & -2[\omega_2^{(1)}][\omega_4^{(1)}] \\
0 & -2[\omega_2^{(1)}][\omega_3^{(1)}] & 1 - 2[\omega_3^{(1)}]^2 & -2[\omega_3^{(1)}][\omega_4^{(1)}] \\
0 & -2[\omega_2^{(1)}][\omega_4^{(1)}] & [\omega_3^{(1)}][\omega_4^{(1)}] & 1 - 2[\omega_4^{(1)}]^2
\end{bmatrix}$$

(9.42)

and the values of $\omega_2^{(1)}$, $\omega_3^{(1)}$, and $\omega_4^{(1)}$ can be calculated from the matrix $\mathbf{A}_0$. The details of these calculations can be found in Wilkinson (1960) and Ralston (1965).

9.5 Eigenvalues of a tridiagonal matrix

Although the above methods effect a considerable simplification in the form of the matrix, this would be of little value unless the resulting tridiagonal matrix was a suitable form for easy solution. In fact, the tridiagonal form leads to a Sturm sequence which is easy to calculate. Thus, approximations to the roots can easily be found and then refined by, for example, the method of bisection.

The Sturm sequence is generated by a recursive sequence as follows. Let $f_r(\lambda)$ be the value of the determinant

$$\begin{vmatrix}
a_1 - \lambda & b_2 & & & \\
b_2 & a_2 - \lambda & b_3 & & 0 \\
& b_3 & a_3 - \lambda & b_4 & \\
0 & & \cdots\cdots\cdots\cdots\cdots & & b_r \\
& & & b_r & a_r - \lambda
\end{vmatrix} = 0$$

(9.43)

which defines the characteristic equation of a tridiagonal matrix. Expanding the determinant by the last column gives

$$f_r(\lambda) = (a_r - \lambda)f_{r-1}(\lambda) - b_r^2 f_{r-2}(\lambda)$$

(9.44)

which is true for $r = n, n - 1, \ldots 2$
If we define

$$\begin{aligned}
f_0(\lambda) &= 1 \\
f_1(\lambda) &= a_1 - \lambda
\end{aligned}$$

(9.45)

then Eqns (9.44) for $r = 2, 3, \ldots, n$ and (9.45) define a sequence of values which are a Sturm sequence. The number of changes in sign of this sequence can be tabulated for various values of λ and the approximate position of the roots determined as described in Chapter 2. Computer results for this method are given in Example 9.6.

A suitable iterative method can then be used to find an accurate value for a root. In this way individual roots or all roots may be found according to the requirements of the problem. The method of bisection for this problem is described in detail in *Modern Computing Methods* (1961). The process is computationally convenient since the process which calculates the Sturm sequence also produces, as one member of the sequence, the value of the function $f_n(\lambda)$ which is required in the iteration equation.

9.6 Further methods

There are two methods which can be used to find all the eigenvalues of a real or complex matrix. These methods are the most efficient available when all the eigenvalues of a matrix must be found. Since, some rather detailed programming is required for these methods, it is not recommended that the user attempts to implement them. Complete routines for these methods have been developed, tested, and published by Wilkinson and others and these standard programs should be used where possible.

The simplest of these methods is the L–R algorithm which was first described by Rutishauser (1958). The name L–R is derived from the computational procedure which involves repeated factorization of a sequence of matrices into left-triangular and right-triangular form. In order to conform with the previous notation the letters **L** and **U** will be used for matrices of lower- and upper-triangular form, respectively.

We form a sequence of matrices by triangular decomposition of each successive member of the sequence. It is assumed for the purpose of the derivation that all the matrices are such that triangular decomposition is possible.

Let

$$\mathbf{A} = \mathbf{A}_1 = \mathbf{L}_1 \mathbf{U}_1 \tag{9.46}$$

and form

$$\mathbf{A}_2 = \mathbf{U}_1 \mathbf{L}_1 = \mathbf{L}_2 \mathbf{U}_2$$
$$\mathbf{A}_r = \mathbf{U}_{r-1} \mathbf{L}_{r-1} = \mathbf{L}_r \mathbf{U}_r, \qquad r = 1, 2, \ldots \tag{9.47}$$

We can see that these matrices are similar matrices to $\mathbf{A}_1$ and therefore have the same eigenvalues, since

$$\mathbf{A}_2 = (\mathbf{U}_1)\mathbf{L}_1 = \mathbf{L}_1^{-1} \mathbf{A}_1 \mathbf{L}_1$$
$$\mathbf{A}_3 = \mathbf{L}_2^{-1} \mathbf{A}_2 \mathbf{L}_2 = \mathbf{L}_2^{-1} \mathbf{L}_1^{-1} \mathbf{A}_1 \mathbf{L}_1 \mathbf{L}_2 \tag{9.48}$$

The sequence of matrices often converge to a block upper-triangular form in which each block corresponds to the eigenvalues of equal modulus. In the case of a matrix with real and distinct eigenvalues the eigenvalues appear in decreasing order from left to right down the diagonal when the matrix has converged. Thus, provided the conditions for triangular decomposition are satisfied, the above method gives a simple repetitive process suitable for computer use. Some numerical results are given in Example 9.7. The eigenvectors are found from the original matrix once the eigenvalues are known. The discussion of the conditions for convergence of the process is beyond the scope of this book. The details are presented at a level suitable for those with some mathematical background in Ralston (1965) and Parlett (1967). A more rigorous presentation is given in Wilkinson (1965).

The second method incorporates orthogonal transformations into the procedure since such transformations have very good stability properties. The matrices are decomposed into a product $Q_r U_r$ where U_r is upper triangular. Thus,

$$A = A_1 = Q_1 U_1 \tag{9.49}$$

$$A_r = U_{r-1} Q_{r-1} = Q_r U_r, \qquad r = 2, 3, \ldots \tag{9.50}$$

We note that, as before, the matrices are similar since

$$A_r = U_{r-1} Q_{r-1} = Q_{r-1}^{-1} A_{r-1} Q_{r-1} \tag{9.51}$$

This method is more complicated and more time-consuming than the L–R method but has the benefit of greater stability. As in the L–R method the matrices converge to upper-triangular form with the eigenvalues on the diagonal. The basic decomposition $A_r = Q_r U_r$ can be achieved for any matrix, which is not the case for the L–U decomposition. When all the eigenvalues of a matrix are required, if the Q–R method from a reliable source is available on a computer this method should be used.

Much of the previous discussion has centred on symmetric matrices which lead to simpler forms for solution and more stable computational schemes. For more general matrices quite severe ill-conditioning may arise. The method suggested for a general matrix is reduction to Hessenberg form (see Eqn (9.39)) by Householder transformation followed by reduction to block-triangular form by the Q–R algorithm.

Bibliographic notes

Some numerical examples showing the calculation of eigenvalues are given in Bull (1966). Detailed mathematical background is not given and the more sophisticated methods are not discussed. A more detailed treatment discussing both computational details and error analysis, which is suitable for the reader with some mathematical background, is given in Fox and Mayers (1968). Two further books which give a good treatment for the reader with some mathematical background and cover the subject in detail are Fox (1964) and Ralston (1965). A rigorous mathematical presentation giving several important theorems and proofs is given in Isaacson and Keller (1966). The reader who wishes to make an exhaustive study of the subject is referred to the text by Wilkinson (1965) which is the standard work on the subject and treats the various aspects of the subject in considerable depth. The papers by Wilkinson in the various journals should also be consulted. In particular the paper in *Computer Journal* (1960) describes Householder's method.

The original paper on the L–R algorithm is due to Rutishauser (1958) and the paper by Francis (1961) describes the Q–R algorithm. The *Linear Algebra Series* of the *Handbook for Automatic Computation* published by Springer Verlag contains programs in ALGOL 60 for the methods described in this chapter. Some of these have been published in *Numerische Mathematik* prior to the publication in book form.

Worked examples

1 Find the two largest latent roots and the corresponding vectors of the matrix below by the power method.

$$\begin{pmatrix} 2\cdot05 & 1\cdot30 & 4\cdot00 \\ 1\cdot30 & 2\cdot15 & 3\cdot70 \\ 4\cdot00 & 3\cdot70 & 8\cdot40 \end{pmatrix}$$

Use the deflation $A_1 = A - \lambda_1 X^{(1)} X^{(1)T}$ assuming that λ_1 is the largest eigenvalue.

Initial vector $[1, 1, 1]$.

The calculation for the first eigenvalue is

x_1		x_2		x_3		λ	
$4\cdot565\,217E$	-1	$4\cdot440\,994E$	-1	$1\cdot000\,000E$	0	$1\cdot610\,000\,00E$	1
$4\cdot644\,941E$	-1	$4\cdot421\,754E$	-1	$1\cdot000\,000E$	0	$1\cdot186\,925\,47E$	1
$4\cdot646\,905E$	-1	$4\cdot417\,781E$	-1	$1\cdot000\,000E$	0	$1\cdot189\,402\,52E$	1
$4\cdot647\,077E$	-1	$4\cdot417\,531E$	-1	$1\cdot000\,000E$	0	$1\cdot189\,334\,09E$	1
$4\cdot647\,089E$	-1	$4\cdot417\,514E$	-1	$1\cdot000\,000E$	0	$1\cdot189\,331\,73E$	1
$4\cdot647\,089E$	-1	$4\cdot417\,512E$	-1	$1\cdot000\,000E$	0	$1\cdot189\,331\,55E$	1
$4\cdot647\,089E$	-1	$4\cdot417\,512E$	-1	$1\cdot000\,000E$	0	$1\cdot189\,331\,53E$	1
$4\cdot647\,089E$	-1	$4\cdot417\,512E$	-1	$1\cdot000\,000E$	0	$1\cdot189\,331\,53E$	1
$1\cdot189\,332E$	1						

Hence, $\lambda_1 = 11\cdot893\,32$ and $X_1^T = [0\cdot464\,708\,9, 0\cdot441\,751\,2, 1\cdot000\,000\,0]$. The deflation process gives the new matrix $A_1 = A - \lambda_1 X^{(1)} X^{(1)T}$.

$2\cdot298\,481E$	-1	$-4\cdot302\,322E$	-1	$8\cdot324\,315E$	-2
$-4\cdot302\,322E$	-1	$5\cdot052\,453E$	-1	$-2\cdot325\,999E$	-2
$8\cdot324\,315E$	-2	$-2\cdot325\,999E$	-2	$-2\cdot840\,871E$	-2

With an initial vector $[1, 1, 1]$ the table of iterations is

x_1		x_2		x_3		λ	
$-1\cdot000\,000E$	0	$4\cdot418\,020E$	-1	$2\cdot695\,424E$	-1	$1\cdot171\,409\,64E$	-1
$-6\cdot141\,836E$	-1	$1\cdot000\,000E$	0	$-1\cdot563\,346E$	-1	$6\cdot471\,810\,00E$	-1
$-7\cdot559\,143E$	-1	$1\cdot000\,000E$	0	$-9\cdot047\,109E$	-2	$7\cdot731\,231\,87E$	-1
$-7\cdot344\,848E$	-1	$1\cdot000\,000E$	0	$-1\cdot004\,296E$	-1	$8\cdot325\,683\,28E$	-1
$-7\cdot375\,264E$	-1	$1\cdot000\,000E$	0	$-9\cdot901\,613E$	-2	$8\cdot235\,802\,71E$	-1
$-7\cdot370\,906E$	-1	$1\cdot000\,000E$	0	$-9\cdot921\,863E$	-2	$8\cdot248\,555\,99E$	-1
$-7\cdot371\,530E$	-1	$1\cdot000\,000E$	0	$-9\cdot918\,966E$	-2	$8\cdot236\,732\,34E$	-1
$-7\cdot371\,441E$	-1	$1\cdot000\,000E$	0	$-9\cdot919\,381E$	-2	$8\cdot246\,993\,83E$	-1
$-7\cdot371\,453E$	-1	$1\cdot000\,000E$	0	$-9\cdot919\,321E$	-2	$8\cdot246\,956\,41E$	-1
$-7\cdot371\,452E$	-1	$1\cdot000\,000E$	0	$-9\cdot919\,330E$	-2	$8\cdot246\,961\,76E$	-1
$-7\cdot371\,452E$	-1	$1\cdot000\,000E$	0	$-9\cdot919\,329E$	-2	$8\cdot246\,961\,00E$	-1
$8\cdot246\,961E$	-1						

Hence, $\lambda_2 = 0\cdot824\,696\,1$ and $[X^{(2)}]^T = [-0\cdot737\,145\,2, 1\cdot000\,000, -0\cdot099\,193\,29]$.

2 Find the eigenvalue nearest to $5\cdot0$ and the corresponding eigenvector of the matrix below by the power method.

$$\begin{pmatrix} 3 & 6 & 8 \\ 4 & 11 & 16 \\ 6 & 15 & 40 \end{pmatrix}$$

The matrix $A - 5I$ is formed and inverse iteration is used. The new matrix is

$$\begin{array}{rrr} -2 & 6 & 8 \\ 4 & 6 & 16 \\ 6 & 15 & 35 \end{array}$$

The solutions of the gaussian elimination are

x_1	x_2	x_3	λ
$7.346\,939E\ -1$	$1.000\,000E\ 0$	$-5.510\,204E\ -1$	$15.465\,986$
$7.462\,063E\ -1$	$1.000\,000E\ 0$	$-5.575\,104E\ -1$	$15.542\,812$
$7.460\,399E\ -1$	$1.000\,000E\ 0$	$-5.574\,888E\ -1$	$15.542\,087$
$7.460\,424E\ -1$	$1.000\,000E\ 0$	$-5.574\,893E\ -1$	$15.542\,098$

The largest eigenvalue of $(\mathbf{A} - 5\mathbf{I})^{-1}$ is, therefore, $15.542\,098$ and the smallest eigenvalue of $\mathbf{A} - 5\mathbf{I}$ is $1/15.542\,098 \sim 0.064$. Hence, $\lambda - 5 = 0.064$ and $\lambda = 5.064$. The eigenvector is $(0.746, 1.00, -0.557)$.

3 The following tables of figures show the results of a calculation to find the root of largest modulus of a matrix, by iteration, for the case where there are two complex conjugate roots of equal modulus. The first nineteen iterations are simply tabulated. The values from iterations 20, 21, and 22 are used to give two simultaneous equations for the coefficients a and b. (See Eqns (9.16).) When these equations are solved the values of a and b are inserted in the quadratic which is solved to give the required eigenvalues. Each consecutive triplet of eigenvectors is used in the same manner and it can be seen that the eigenvalues have converged. There are different values for the eigenvectors at each stage because the eigenvector can be arbitrarily multiplied by a scalar multiplier. The eigenvectors are $[i, -1/\sqrt{2}, -1/\sqrt{2}, 0]$ and $[-i, -1/\sqrt{2}, -1/\sqrt{2}, 0]$. The eigenvectors are each tabulated in two columns, the first being the real parts and the second the imaginary parts of the vectors.

The calculation took 2.60 seconds on an ICL 1904A computer.

A

$3.000\,000\,000\,0$	$-1.414\,213\,562\,4$	$1.414\,213\,562\,4$	$0.000\,000\,000\,0$
$1.414\,213\,562\,4$	$2.250\,000\,000\,0$	$-0.750\,000\,000\,0$	$-0.353\,553\,390\,6$
$-1.414\,213\,562\,4$	$-0.750\,000\,000\,0$	$2.250\,000\,000\,0$	$-0.353\,553\,390\,6$
$0.000\,000\,000\,0$	$-0.353\,553\,390\,6$	$-0.353\,553\,390\,6$	$1.060\,660\,171\,8$

$\mathbf{Y}^{(0)}$

$1.000\,000\,000\,0$
$1.000\,000\,000\,0$
$1.000\,000\,000\,0$
$1.000\,000\,000\,0$

$\mathbf{Y}^{(1)}$

$1.000\,000\,000\,0$
$0.853\,553\,390\,6$
$-0.089\,255\,651\,0$
$0.117\,851\,130\,2$

$\mathbf{Y}^{(2)}$

$0.496\,034\,143\,1$
$1.000\,000\,000\,0$
$-0.683\,595\,710\,2$
$-0.043\,220\,464\,0$

$\mathbf{Y}^{(3)}$

$-0.256\,608\,042\,4$
$1.000\,000\,000\,0$
$-0.854\,814\,933\,0$
$-0.045\,325\,228\,9$

$\mathbf{Y}^{(4)}$

$-1.000\,000\,000\,0$
$0.749\,864\,746\,1$
$-0.676\,232\,963\,8$
$-0.029\,297\,798\,6$

$Y^{(20)}$	$Y^{(21)}$	$Y^{(22)}$
1·000 000 000 0	5·085 482 313 8	17·512 893 882 6
−0·737 328 667 9	−0·797 773 233 7	4·798 636 911 2
0·737 330 018 2	0·797 775 699 9	−4·798 632 406 8
−0·000 000 623 4	−0·000 001 138 6	−0·000 002 079 6

A,B = −9·6114, 16·9074

EIGENVALUES = 7·293 075 +i* 0·000 000, 2·318 283 +i* 0·000 000

Eigenvector		*Eigenvector*	
−0·556 244 160 4	0·000 000 000 0	0·443 755 839 6	0·000 000 000 0
−0·183 236 503 8	0·000 000 000 0	−0·920 565 171 7	0·000 000 000 0
0·183 236 637 2	0·000 000 000 0	0·920 566 655 4	0·000 000 000 0
−0·000 000 061 6	0·000 000 000 0	−0·000 000 685 0	0·000 000 000 0

$Y^{(23)}$	$Y^{(24)}$	$Y^{(25)}$
0·994 975 167 3	0·156 498 674 4	−11·995 685 128 9
1·000 000 000 0	4·407 107 252 6	13·442 644 080 4
−0·999 999 789 9	−4·407 106 868 9	−13·442 643 379 6
−0·000 000 097 0	−0·000 000 177 1	−0·000 000 323 6

A,B = −6·0000, 13·0000

EIGENVALUES = 3·000 000 +i* 2·000 000, 3·000 000 +i* −2·000 00

Eigenvector		*Eigenvector*	
−0·497 487 583 7	−0·707 106 756 8	0·497 487 583 7	−0·707 106 756 8
−0·500 000 000 0	0·351 776 813 2	0·500 000 000 0	0·351 776 813 2
0·499 999 895 0	−0·351 776 874 8	−0·499 999 895 0	−0·351 776 874 8
0·000 000 048 5	0·000 000 028 5	−0·000 000 048 5	0·000 000 028 5

$Y^{(26)}$	$Y^{(27)}$	$Y^{(28)}$
−1·000 000 000 0	−3·892 894 605 7	−10·357 367 634 2
0·315 685 923 9	−0·467 155 800 7	−6·906 851 769 0
−0·315 685 906 6	0·467 155 832 3	6·906 851 826 8
−0·000 000 008 0	−0·000 000 014 6	−0·000 000 026 6

A,B = −6·0000, 13·0000

EIGENVALUES = 3·000 000 +i* 2·000 000, 3·000 000 +i* −2·000 000

Eigenvector		*Eigenvector*	
0·500 000 000 0	−0·223 223 655 5	−0·500 000 000 0	−0·223 223 655 5
−0·157 842 962 0	−0·353 553 393 1	0·157 842 962 0	−0·353 553 393 1
0·157 842 953 3	0·353 553 388 1	−0·157 842 953 3	0·353 553 388 1
0·000 000 004 0	0·000 000 002 3	−0·000 000 004 0	0·000 000 002 3

$Y^{(29)}$	$Y^{(30)}$	$Y^{(31)}$
−0·326 186 047 6	1·849 868 977 7	15·339 632 485 1
−0·999 999 997 0	−3·461 296 725 1	−7·767 780 381 8
1·000 000 000 0	3·461 296 730 7	7·767 780 391 9
−0·000 000 001 4	−0·000 000 002 5	−0·000 000 004 6

A,B = −6·0000, 13·0000

EIGENVALUES = 3·000 000 +i* 2·000 000, 3·000 000 +i* −2·000 000

Eigenvector		*Eigenvector*	
0·163 093 023 8	0·707 106 779 4	−0·163 093 023 8	0·707 106 779 4
0·499 999 998 5	−0·115 324 183 5	−0·499 999 998 5	−0·115 324 183 5
−0·500 000 000 0	0·115 324 182 7	0·500 000 000 0	0·115 324 182 7
0·000 000 000 7	0·000 000 000 4	−0·000 000 000 7	0·000 000 000 4

$Y^{(32)}$	$Y^{(33)}$	$Y^{(34)}$
1·000 000 000 0	3·066 970 229 5	5·401 821 376 8
−0·023 677 551 6	1·343 180 907 6	8·366 893 616 3
0·023 677 551 8	−1·343 180 907 1	−8·366 893 615 3
−0·000 000 000 1	−0·000 000 000 2	−0·000 000 000 4

A, B = −6·0000, 13·0000

EIGENVALUES = 3·000 000 $+i*$ 2·000 000, 3·000 000 $+i*$ −2·000 000

Eigenvector		*Eigenvector*	
−0·500 000 000 0	0·016 742 557 3	0·500 000 000 0	0·016 742 557 3
0·011 838 775 8	0·353 553 390 5	−0·011 838 775 8	0·353 553 390 5
−0·011 838 775 9	−0·353 553 390 6	0·011 838 775 9	−0·353 553 390 6
0·000 000 000 1	0·000 000 000 0	−0·000 000 000 1	0·000 000 000 0

4 The Jacobi method uses a similarity transformation to reduce an off-diagonal element to zero. In the case of a symmetric matrix the symmetric element is also reduced to zero. The off-diagonal elements are systematically eliminated until the off-diagonal elements are effectively zero. The diagonal elements then give the approximations to the eigenvalues. The most effective system for eliminating off-diagonal elements is to set a threshold value and systematically scan the rows eliminating all elements with modulus greater than the threshold value. The threshold value is then reduced gradually until an acceptable final level is reached. In the computer results presented below, the matrix **A** represents the initial matrix and its subsequent transformation, the matrix **C** represents the matrix used in the similarity transform $C_r^T A_{r-1} C_r$, and the final eigenvectors are the columns of the matrix **CE** which is formed as the product $C_1 C_2 \cdots C_r$. The latter result arose because the eigenvectors of the final diagonal matrix are the columns of the unit matrix and, using the theory previously presented, the eigenvectors of the original matrix are the columns of $C_1 C_2 \cdots C_r I$. There are 16 steps in the computation presented and the threshold value is reduced to 10^{-5}. In order to reduce the volume of numbers presented the first few and the last few transformations only are presented.

The calculation took 2·66 seconds on an ICL 1904A computer.

A

10·000 000 000 0	7·000 000 000 1	4·000 000 000 0	1·000 000 000 0
7·000 000 000 1	11·000 000 000 0	1·000 000 000 0	2·000 000 000 0
4·000 000 000 0	1·000 000 000 0	5·000 000 000 0	3·000 000 000 0
1·000 000 000 0	2·000 000 000 0	3·000 000 000 0	4·000 000 000 0

THRES = $0·10E$ 01
1   THETA = −0·749 744 $I = 1$ $J = 2$

C

0·731 863 050 7	0·681 451 740 8	0·000 000 000 0	0·000 000 000 0
−0·681 451 740 8	0·731 863 050 7	0·000 000 000 0	0·000 000 000 0
0·000 000 000 0	0·000 000 000 0	1·000 000 000 0	0·000 000 000 0
0·000 000 000 0	0·000 000 000 0	0·000 000 000 0	1·000 000 000 0

A

3·482 165 576 3	0·000 000 000 0	2·246 000 462 1	−0·631 040 430 9
−0·000 000 000 1	17·517 834 424 5	3·457 670 013 9	2·145 177 842 2
2·246 000 462 1	3·457 670 013 9	5·000 000 000 0	3·000 000 000 0
−0·631 040 430 9	2·145 177 842 2	3·000 000 000 0	4·000 000 000 0

CE

0·731 863 050 7	0·681 451 740 8	0·000 000 000 0	0·000 000 000 0
−0·681 451 740 8	0·731 863 050 7	0·000 000 000 0	0·000 000 000 0
0·000 000 000 0	0·000 000 000 0	1·000 000 000 0	0·000 000 000 0
0·000 000 000 0	0·000 000 000 0	0·000 000 000 0	1·000 000 000 0

2   THETA $= -0.622472$ $I = 1$ $J = 3$

C

0·812 439 666 0	0·000 000 000 0	0·583 045 271 9	0·000 000 000 0
0·000 000 000 0	1·000 000 000 0	0·000 000 000 0	0·000 000 000 0
−0·583 045 271 9	0·000 000 000 0	0·812 439 666 0	0·000 000 000 0
0·000 000 000 0	0·000 000 000 0	0·000 000 000 0	1·000 000 000 0

A

1·870 329 024 1	−2·015 978 153 5	−0·000 000 000 0	−2·261 818 092 7
−2·015 978 153 6	17·517 834 424 5	2·809 148 271 2	2·145 177 842 2
−0·000 000 000 0	2·809 148 271 3	6·611 836 552 6	2·069 393 858 4
−2·261 818 092 7	2·145 177 842 2	2·069 393 858 4	4·000 000 000 0

CE

0·594 594 572 5	0·681 451 740 8	0·426 709 291 4	0·000 000 000 0
−0·553 638 424 7	0·731 863 050 7	−0·397 317 215 5	0·000 000 000 0
−0·583 045 271 9	0·000 000 000 0	0·812 439 666 0	0·000 000 000 0
0·000 000 000 0	0·000 000 000 0	0·000 000 000 0	1·000 000 000 0

3   THETA $= 0.565395$ $I = 1$ $J = 4$

C

0·844 376 860 3	0·000 000 000 0	0·000 000 000 0	−0·535 749 678 3
0·000 000 000 0	1·000 000 000 0	0·000 000 000 0	0·000 000 000 0
0·000 000 000 0	0·000 000 000 0	1·000 000 000 0	0·000 000 000 0
0·535 749 678 3	0·000 000 000 0	0·000 000 000 0	0·844 376 860 3

A

0·435 225 372 6	−0·552 966 964 9	1·108 677 093 8	0·000 000 000 1
−0·552 966 964 9	17·517 834 424 5	2·809 148 271 2	2·891 398 178 4
1·108 677 093 9	2·809 148 271 3	6·611 836 552 6	1·747 348 288 9
0·000 000 000 0	2·891 398 178 4	1·747 348 288 9	5·435 103 651 5

CE

0·502 061 898 3	0·681 451 740 8	0·426 709 291 4	−0·318 553 850 9
−0·467 479 474 8	0·731 863 050 7	−0·397 317 215 5	0·296 611 607 9
−0·492 309 936 1	0·000 000 000 0	0·812 439 666 0	0·312 366 316 9
0·535 749 678 3	0·000 000 000 0	0·000 000 000 0	0·844 376 860 3

4   THETA $= 0.237850$ $I = 2$ $J = 3$

C

1·000 000 000 0	0·000 000 000 0	0·000 000 000 0	0·000 000 000 0
0·000 000 000 0	0·971 846 882 0	−0·235 613 322 9	0·000 000 000 0
0·000 000 000 0	0·235 613 322 9	0·971 846 882 0	0·000 000 000 0
0·000 000 000 0	0·000 000 000 0	0·000 000 000 0	1·000 000 000 0

A

0·435 225 372 6	−0·276 180 126 6	1·207 750 760 8	0·000 000 000 1
−0·276 180 126 6	18·198 880 761 1	0·000 000 000 0	3·221 694 840 9
1·207 750 760 9	−0·000 000 000 1	5·930 790 216 0	1·016 903 053 7
0·000 000 000 0	3·221 694 840 9	1·016 903 053 7	5·435 103 651 5

CE

0·502 061 898 3	0·762 805 143 6	0·254 136 985 4	−0·318 553 850 9
−0·467 479 474 8	0·617 645 594 5	−0·558 568 182 4	0·296 611 607 9
−0·492 309 936 1	0·191 421 609 4	0·789 566 956 2	0·312 366 316 9
0·535 749 678 3	0·000 000 000 0	0·000 000 000 0	0·844 376 860 3

5   THETA = 0·233 747 $I = 2$ $J = 4$

C

1·000 000 000 0	0·000 000 000 0	0·000 000 000 0	0·000 000 000 0
0·000 000 000 0	0·972 805 225 4	0·000 000 000 0	−0·231 624 682 2
0·000 000 000 0	0·000 000 000 0	1·000 000 000 0	0·000 000 000 0
0·000 000 000 0	0·231 624 682 2	0·000 000 000 0	0·972 805 225 4

A

0·435 225 372 6	−0·268 669 470 3	1·207 750 760 8	0·063 970 134 1
−0·268 669 470 3	18·965 965 501 3	0·235 539 846 7	−0·000 000 000 0
1·207 750 760 9	0·235 539 846 6	5·930 790 216 0	0·989 248 604 4
0·063 970 134 1	−0·000 000 000 1	0·989 248 604 4	4·668 018 911 2

CE

0·502 061 898 3	0·668 275 895 2	0·254 136 985 4	−0·486 575 349 7
−0·467 479 474 8	0·669 551 431 2	−0·558 568 182 4	0·145 483 357 5
−0·492 309 936 1	0·258 567 690 7	0·789 566 956 2	0·259 533 615 9
0·535 749 678 3	0·195 578 521 9	0·000 000 000 0	0·821 414 221 9

6   THETA = 0·207 059 $I = 3$ $J = 1$

C

0·978 639 711 7	0·000 000 000 0	0·205 582 865 7	0·000 000 000 0
0·000 000 000 0	1·000 000 000 0	0·000 000 000 0	0·000 000 000 0
−0·205 582 865 7	0·000 000 000 0	0·978 639 711 7	0·000 000 000 0
0·000 000 000 0	0·000 000 000 0	0·000 000 000 0	1·000 000 000 0

A

0·181 513 143 7	−0·311 353 569 6	−0·000 000 000 1	−0·140 768 849 4
−0·311 353 569 7	18·965 965 501 3	0·175 274 808 0	−0·000 000 000 0
−0·000 000 000 0	0·175 274 807 9	6·184 502 444 8	0·981 269 132 5
−0·140 768 849 4	−0·000 000 000 1	0·981 269 132 5	4·668 018 911 2

CE

0·439 091 501 6	0·668 275 895 2	0·351 923 869 9	−0·486 575 349 7
−0·342 661 930 8	0·669 551 431 2	−0·642 742 775 1	0·145 483 357 5
−0·644 115 491 5	0·258 567 690 7	0·671 491 090 9	0·259 533 615 9
0·524 305 910 7	0·195 578 521 9	0·110 140 954 2	0·821 414 221 9

THRES = 0·10E 00

7   THETA = 0·016 569 $I = 1$ $J = 2$

C

0·999 862 737 3	−0·016 568 242 0	0·000 000 000 0	0·000 000 000 0
0·016 568 242 0	0·999 862 737 3	0·000 000 000 0	0·000 000 000 0
0·000 000 000 0	0·000 000 000 0	1·000 000 000 0	0·000 000 000 0
0·000 000 000 0	0·000 000 000 0	0·000 000 000 0	1·000 000 000 0

A

0·176 353 854 3	0·000 000 000 0	0·002 903 995 4	−0·140 749 527 1
−0·000 000 000 1	18·971 124 790 9	0·175 250 749 3	0·002 332 292 3
0·002 903 995 4	0·175 250 749 2	6·184 502 444 8	0·981 269 132 5
−0·140 749 527 1	0·002 332 292 3	0·981 269 132 5	4·668 018 911 2

CE

0·450 103 387 5	0·660 909 191 6	0·351 923 869 9	−0·486 575 349 7
−0·331 521 605 9	0·675 136 832 5	−0·642 742 775 1	0·145 483 357 5
−0·639 743 066 4	0·269 204 060 3	0·671 491 090 9	0·259 533 615 9
0·527 474 335 3	0·186 864 849 1	0·110 140 954 2	0·821 414 221 9

THRES $= 0 \cdot 10E \quad -03$
15   THETA $= -0 \cdot 000\,029 \quad I = 1 \quad J = 2$

C

0·999 999 999 6	0·000 029 430 4	0·000 000 000 0	0·000 000 000 0
−0·000 029 430 4	0·999 999 999 6	0·000 000 000 0	0·000 000 000 0
0·000 000 000 0	0·000 000 000 0	1·000 000 000 0	0·000 000 000 0
0·000 000 000 0	0·000 000 000 0	0·000 000 000 0	1·000 000 000 0

A

0·171 752 891 4	0·000 000 000 0	−0·000 000 567 6	−0·000 000 529 5
−0·000 000 000 1	18·973 543 783 6	−0·000 000 000 0	−0·000 000 042 8
−0·000 000 567 6	−0·000 000 000 1	6·664 858 254 3	−0·000 075 562 4
−0·000 000 529 5	−0·000 000 042 8	−0·000 075 562 4	4·189 845 072 8

CE

0·432 011 697 6	0·665 155 644 9	0·088 255 154 5	−0·602 615 042 1
−0·322 845 225 7	0·666 374 137 9	−0·518 019 542 0	0·428 219 829 9
−0·634 823 972 1	0·278 727 841 8	0·719 401 054 4	−0·042 088 450 4
0·553 295 018 9	0·189 272 994 8	0·454 234 416 0	0·672 094 822 8

THRES $= 0 \cdot 10E \quad -04$
16   THETA $= -0 \cdot 000\,031 \quad I = 3 \quad J = 4$

C

1·000 000 000 0	0·000 000 000 0	0·000 000 000 0	0·000 000 000 0
0·000 000 000 0	1·000 000 000 0	0·000 000 000 0	0·000 000 000 0
0·000 000 000 0	0·000 000 000 0	0·999 999 999 5	0·000 030 530 1
0·000 000 000 0	0·000 000 000 0	−0·000 030 530 1	0·999 999 999 5

A

0·171 752 891 4	0·000 000 000 0	−0·000 000 567 6	−0·000 000 529 5
−0·000 000 000 1	18·973 543 783 6	−0·000 000 000 0	−0·000 000 042 8
−0·000 000 567 5	−0·000 000 000 1	6·664 858 256 7	0·000 000 000 0
−0·000 000 529 5	−0·000 000 042 8	0·000 000 000 0	4·189 845 070 4

CE

0·432 011 697 6	0·665 155 644 9	0·088 273 552 3	−0·602 612 347 3
−0·322 845 225 7	0·666 374 137 9	−0·518 032 615 4	0·428 204 014 5
−0·634 823 972 1	0·278 727 841 8	0·719 402 339 1	−0·042 066 487 0
0·553 295 018 9	0·189 272 994 8	0·454 213 896 7	0·672 108 690 3

THRES $= 0 \cdot 10E \quad -05$

5 The figures below are the results of a calculation to transform the matrix A_0 into a tridiagonal matrix by the Givens' method. The transformation takes the form $A_r = P_r^T A_{r-1} P_r$ where the matrix P_r is a rotation matrix as discussed in Section 9.4.2. For example, the first element to be eliminated from the matrix A_0 is element 1, 3 which is achieved using a rotation matrix rotating in the (2, 3) plane. The angle θ is found from the formula

$$\tan \theta = \frac{a_{13}}{a_{12}} = 3$$

Therefore,

$$\sin \theta = \frac{3}{\sqrt{10}} = 0 \cdot 948\,683, \quad \cos \theta = \frac{1}{\sqrt{10}} = 0 \cdot 316\,228$$

The eigenvalues of the tridiagonal matrix can then be found by the Sturm sequence method. (See Example 9.6.)

$$A_0 = \begin{bmatrix} 2.000\,000 & 1.000\,000 & 3.000\,000 & 2.000\,000 \\ 1.000\,000 & 4.000\,000 & 2.000\,000 & 1.000\,000 \\ 3.000\,000 & 2.000\,000 & 3.000\,000 & -3.000\,000 \\ 2.000\,000 & 1.000\,000 & -3.000\,000 & 1.000\,000 \end{bmatrix}$$

$$P_1 = \begin{bmatrix} 1.000\,000 & 0.000\,000 & 0.000\,000 & 0.000\,000 \\ 0.000\,000 & 0.316\,228 & -0.948\,683 & 0.000\,000 \\ 0.000\,000 & 0.948\,683 & 0.316\,228 & 0.000\,000 \\ 0.000\,000 & 0.000\,000 & 0.000\,000 & 1.000\,000 \end{bmatrix}$$

$$A_1 = \begin{bmatrix} 2.000\,000 & 3.162\,278 & -0.000\,000 & 2.000\,000 \\ 3.162\,278 & 4.300\,000 & -1.900\,000 & -2.529\,822 \\ -0.000\,000 & -1.900\,000 & 2.700\,000 & -1.897\,367 \\ 2.000\,000 & -2.529\,822 & -1.897\,367 & 1.000\,000 \end{bmatrix}$$

$$P_2 = \begin{bmatrix} 1.000\,000 & 0.000\,000 & 0.000\,000 & 0.000\,000 \\ 0.000\,000 & 0.845\,154 & 0.000\,000 & -0.534\,522 \\ 0.000\,000 & 0.000\,000 & 1.000\,000 & 0.000\,000 \\ 0.000\,000 & 0.534\,522 & 0.000\,000 & 0.845\,154 \end{bmatrix}$$

$$A_2 = \begin{bmatrix} 2.000\,000 & 3.741\,657 & -0.000\,000 & -0.000\,000 \\ 3.741\,657 & 1.071\,429 & -2.619\,978 & -2.574\,998 \\ -0.000\,000 & -2.619\,978 & 2.700\,000 & -0.587\,975 \\ -0.000\,000 & -2.574\,998 & -0.587\,975 & 4.228\,571 \end{bmatrix}$$

$$P_3 = \begin{bmatrix} 1.000\,000 & 0.000\,000 & 0.000\,000 & 0.000\,000 \\ 0.000\,000 & 1.000\,000 & 0.000\,000 & 0.000\,000 \\ 0.000\,000 & 0.000\,000 & 0.713\,203 & -0.700\,958 \\ 0.000\,000 & 0.000\,000 & 0.700\,958 & 0.713\,203 \end{bmatrix}$$

$$A_3 = \begin{bmatrix} 2.000\,000 & 3.741\,657 & -0.000\,000 & 0.000\,000 \\ 3.741\,657 & 1.071\,429 & -3.673\,540 & 0.000\,000 \\ -0.000\,000 & -3.673\,540 & 2.863\,165 & 0.753\,990 \\ 0.000\,000 & 0.000\,000 & 0.753\,990 & 4.065\,406 \end{bmatrix}$$

6 The following table shows the Sturm sequence of calculations at a series of points when finding approximations to the eigenvalues.

$$\begin{bmatrix} 1 & -2 & 0 & 0 \\ -2 & 1 & -2 & 0 \\ 0 & -2 & 1 & -2 \\ 0 & 0 & -2 & 1 \end{bmatrix}$$

λ	-10	-2	0	2	10
f_0	1	1	1	1	1
f_1	11	3	1	-1	-9
f_2	117	5	-3	-3	77
f_3	1243	3	-7	7	-657
f_4	11962	-14	12	-2	6262
Number of sign changes	0	1	2	3	4

Hence, the roots lie between -10 and -2, -2, and $0,0$, and $2,2$ and 10. Thus, approximate positions have been found for all the roots.

7 A table of calculations of the L–R method is given below. The true solutions for the three eigenvalues are $8.449\,490$, $6.000\,000$, and $3.550\,510$. The matrices in the left column are made up of the complete upper-triangular matrix coefficients together with the lower-triangular matrix coefficients without the diagonal elements which are unity. The right-hand column give the new matrices formed by the product $\mathbf{U}_r\mathbf{L}_r$. Thus, in the first case the matrix is factorized as follows.

$$\begin{bmatrix} 8 & 1 & 0 \\ 1 & 6 & 1 \\ 0 & 1 & 4 \end{bmatrix} = \begin{bmatrix} 1 & 0 & 0 \\ 0{\cdot}125 & 1 & 0 \\ 0 & 0{\cdot}170\,213 & 1 \end{bmatrix}\begin{bmatrix} 8 & 1 & 0 \\ 0 & 5{\cdot}875 & 0 \\ 0 & 0 & 3{\cdot}829\,787 \end{bmatrix}$$

The new matrix is found as follows.

$$\begin{bmatrix} 8 & 1 & 0 \\ 0 & 5{\cdot}875 & 0 \\ 0 & 0 & 3{\cdot}829\,787 \end{bmatrix}\begin{bmatrix} 1 & 0 & 0 \\ 0{\cdot}125 & 1 & 1 \\ 0 & 0{\cdot}170\,213 & 1 \end{bmatrix}$$
$$= \begin{bmatrix} 8{\cdot}125 & 1 & 0 \\ 0{\cdot}734\,375 & 6{\cdot}045\,230 & 0 \\ 0 & 0{\cdot}651\,880 & 3{\cdot}829\,787 \end{bmatrix}$$

It can be seen that convergence is slow but the lower-triangular elements are becoming insignificant and the diagonal elements are approaching the true values of the eigenvalues.

8·0	1·000000	0	8·125	1·000000	0
0·125	5·875	1·000000	0·734375	6·045230	1·000000
0	0·170213	3·829787	0	0·651880	3·829787

8·125	1·000000	0	8·215385	1·000000	0
0·090385	5·954845	1·000000	0·538229	6·062679	1·000000
0	0·107834	3·721953	0	0·401353	3·721953

8·215385	1·000000	0	8·280900	1·000000	0
0·065515	5·997164	1·000000	0·392904	6·064088	1·000000
0	0·066924	3·655029	0	0·244609	3·655029

8·280900	1·000000	0	8·328347	1·000000	0
0·047447	6·016641	1·000000	0·285472	6·057296	1·000000
0	0·040655	3·614374	0	0·146942	3·614374

8·328347	1·000000	0	8·362624	1·000000	0
0·034277	6·023019	1·000000	0·206451	6·047416	1·000000
0	0·024397	3·589977	0	0·087585	3·589977

8·362624	1·000000	0	8·387311	1·000000	0
0·024687	6·022729	1·000000	0·148683	6·037271	1·000000
0	0·014542	3·575435	0	0·051994	3·575435

8·387311	1·000000	0	8·405038	1·000000	0
0·017727	6 019544	1 000000	0·106708	6·028182	1·000000
0	0·008638	3·566797	0	0·030810	3·566797

8·405038	1·000000	0	8·417734	1·000000	0
0·012696	6·015486	1·000000	0·076373	6·020608	1·000000
0	0·005122	3·561675	0	0·018243	3·561675

8·417734	1·000000	0	8·426807	1·000000	0
0·009073	6·011535	1·000000	0·054543	6·014570	1·000000
0	0·003035	3·558640	0	0·010800	3·558640

8·426807	1·000000	0	8·433280	1·000000	0
0·006473	6·008097	1·000000	0·038890	6·009895	1·000000
0	0·001798	3·556842	0	0·006395	3·556842

8·433280	1·000000	0	8·437891	1·000000	0
0·004611	6·005284	1·000000	0·027690	6·006349	1·000000
0	0·001065	3·555777	0	0·003787	3·555777

8·437891	1·000000	0	8·441173	1·000000	0
0·003282	6·003067	1·000000	0·019702	6·003698	1·000000
0	0·000631	3·555146	0	0·002243	3·555146
8·441173	1·000000	0	8·443507	1·000000	0
0·002334	6·001364	1·000000	0·014007	6·001738	1·000000
0	0·000374	3·554772	0	0·001329	3·554772
8·443507	1·000000	0	8·445166	1·000000	0
0·001659	6·000079	1·000000	0·009954	6·000301	1·000000
0	0·000222	3·554550	0	0·000789	3·554550

Problems

1 Find the Jacobi transformation matrix which will produce zero in position $(1,3)$ of the transformed matrix for the following matrix. Complete the reduction to diagonal form and find all the eigenvalues and eigenvectors.

$$\begin{bmatrix} 7 & 0 & 6 \\ 0 & 5 & 0 \\ 6 & 0 & 2 \end{bmatrix}$$

2 Find the eigenvalue of largest modulus and the corresponding normalized eigenvector of the matrix

$$\begin{bmatrix} 3 & 6 & 8 \\ 4 & 11 & 16 \\ 6 & 15 & 40 \end{bmatrix}$$

3 Find the eigenvalue nearest in value to $4\cdot0$ of the matrix

$$\begin{bmatrix} 1 & 2 & 0 \\ 4 & 3 & 0 \\ 5 & 6 & 7 \end{bmatrix}$$

Find also the corresponding eigenvector.

4 Find the eigenvalue of largest modulus of the matrix

$$\begin{bmatrix} 0 & \sqrt{2} & 1 \\ -\sqrt{2} & 0 & 1 \\ -1 & -1 & 0 \end{bmatrix}$$

and find also the corresponding eigenvector.

5 Find all the eigenvalues and eigenvectors of the matrix

$$\begin{bmatrix} 2 & -1 & 0 \\ -1 & 2 & -1 \\ 0 & -1 & 2 \end{bmatrix}$$

6 Reduce the following matrix to tridiagonal form using the Givens transformation. Use the Sturm sequence method in conjunction with the bisection technique to find the eigenvalues of the tridiagonal matrix. Find also the eigenvectors corresponding to all the eigenvalues

$$\begin{bmatrix} 1 & 1 & 1 & 1 \\ 1 & 2 & 3 & 4 \\ 1 & 3 & 6 & 10 \\ 1 & 4 & 10 & 20 \end{bmatrix}$$

Bibliography

Balfour, A. and McTernan, A. J., 1967, *The Numerical Solution of Equations*, Heinemann, London.

Bareiss, E. H., 1967, The numerical solution of polynomial equations and the resultant procedures. In Ralston and Wilf, 1967.

Bauer, F. L., 1963, Optimally scaled matrices, *Num. Math.*, **5**, 73–87.

Buckingham, R. A., 1957, *Numerical Methods*, Pitman, London.

Bull, G., 1966, *Computational Methods and Algol*, Harrap, London.

Butler, R. and Kerr, E., 1962, *An Introduction to Numerical Methods*, Pitman, London.

Conte, S. D., 1965, *Elementary Numerical Analysis*, McGraw-Hill, New York.

Dahlquist, G., 1956, Convergence and stability in the numerical integration of ordinary differential equations, *Math. Scand.*, **4**, 33–53.

Davis, P. J., 1964, *Interpolation and Approximation*, Blaisdell, New York.

Davis, P. J. and Rabinowitz, P., 1967, *Numerical Integration*, Blaisdell, Waltham, Mass.

Forsythe, G. E. and Moler, C. B., 1967, *Computer Solution of Linear Algebraic Systems*, Prentice-Hall, Englewood Cliffs, N.J.

Fox, L. (Ed.), 1962, *Numerical Solution of Ordinary and Partial Differential Equations*, Pergamon Press, Oxford.

Fox, L., 1964, *An Introduction to Numerical Linear Algebra*, Clarendon Press, Oxford.

Fox, L. and Mayers, D. F., 1968, *Computing Methods for Scientists and Engineers*, Clarendon Press, Oxford.

Francis, J. G. F., 1961, The Q–R transformation—a unitary analogue to the L–R transformation, *Comp. J.*, **4**, 265–271, 332–345.

Gear, C. W., 1968, The automatic integration of stiff ordinary differential equations, *Proc. IFIPS Conf.*, Edinburgh.

Goldberg, S., 1958, *Introduction to Difference Equations*, John Wiley, New York.

Hamming, R. W., 1962, *Numerical Methods for Scientists and Engineers*, McGraw-Hill, New York.

Hartree, D. R., 1958, *Numerical Analysis*, Oxford University Press, London.

Henrici, P., 1962, *Discrete Variable Methods in Ordinary Differential Equations*, John Wiley, New York.

Henrici, P., 1963, *Error Propagation for Difference Methods*, John Wiley, New York.

Henrici, P., 1964, *Elements of Numerical Analysis*, John Wiley, New York.

Hildebrand, F. B., 1956, *Introduction to Numerical Analysis*, McGraw-Hill, New York.

Householder, A. S., 1953, *Principles of Numerical Analysis*, McGraw-Hill, New York.

Isaacson, E. and Keller, H. B., 1966, *Analysis of Numerical Methods*, John Wiley, New York.

Keller, H. B., 1968, *The Numerical Solution of Boundary Value Problems*, Blaisdell, New York.

Krylov, V. I., 1962, *Approximate Calculation of Integrals*, trans. Stroud, A. H., Macmillan, New York.

Lanczos, C., 1957, *Applied Analysis*, Pitman, London.

Lehmer, D. H., 1961, A machine method for solving polynomial equations, *J.A.C.M.*, **8**, 151–162.

McCracken, D. D., and Dorn, W. S., 1964, *Numerical Methods and* FORTRAN *Programming*, John Wiley, New York.

Modern Computing Methods, 1961, HMSO, London.

Nordsieck, A., 1962, Numerical integration of ordinary differential equations, *Maths Comp.*, **16**, 22–49.

Ostrowski, A. M., 1960, *Solution of Equations and Systems of Equations*, Academic Press, New York.

Parlett, B. N., 1967, The L–U and Q–R algorithms. In Ralston and Wilf, 1967.

Ralston, A., 1965, *A First Course in Numerical Analysis*, McGraw-Hill, New York.

Ralston, A. and Wilf, H. S., 1960, *Mathematical Methods for Digital Computers*, Vol. 1, John Wiley, New York.

Ralston, A. and Wilf, H. S., 1967, *Mathematical Methods for Digital Computers*, Vol. 2, John Wiley, New York.

Redish, K. A., 1961, *An Introduction to Computational Methods*, English Universities Press, London.

Rice, J. R., 1964, *The Approximation of Functions*, Vol. 1, Addison-Wesley, New York.

Rutishauser, H. 1956, Der Quotienten–Differenzen–Algorithmus. *Mitteilungen aus dem Institut für angew. Math. No. 7*, Birkhauser, Basel and Stuttgart.

Rutishauser, H., 1958, Solution of eigenvalue problems with the L–R transformation, *App. Math. Ser. Nat. Bur. Stand.*, **49**, 47–81.

Scarborough, J. B., 1958, *Numerical Mathematical Analysis*, Johns Hopkins Press, Baltimore.

Smith, G. D., 1964, *The Numerical Solution of Partial Differential Equations*, Oxford University Press, London.

Stroud, A. H., and Secrest, D., 1966, *Gaussian Quadrature Formulae*, Prentice-Hall, Englewood Cliffs, N.J.

Todd, J., 1962, *Survey of Numerical Analysis*, McGraw-Hill, New York.

Traub, J. F., 1964, *Iterative Methods for the Solution of Equations*, Prentice-Hall, Englewood Cliffs, N.J.

Varga, R. S., 1962, *Matrix Iterative Analysis*, Prentice-Hall, Englewood Cliffs, N.J.

Wilkinson, J. H., 1960, Householder's method for the solution of the algebraic eigenproblem, *Comp. J.*, **3**, 23–27.

Wilkinson, J. H., 1961, Error analysis of direct methods of matrix inversion, *J.A.C.M.*, **8**, 281–330.

Wilkinson, J. H., 1963, *Rounding Errors In Algebraic Processes*, HMSO, London.

Wilkinson, J. H., 1965, *The Algebraic Eigenvalue Problem*, Oxford University Press, London.

Answers to problems

Chapter 1

1 100 101, 111 101, 11 010 011, 1 101 011, 111 001, 1 111 111

2 13, 30, 86, 219, 85, 204

3 $10 111 + 10 001 = 101 000(40)$, $10 010 - 1100 = 110(6)$,
$1111 \times 1110 = 11 010 010(210)$, $111 111 \div 1001 = 111(7)$

4 $\dfrac{e_s}{s} = e_1\left(\dfrac{x_1 + x_2}{x_1 + x_2 + x_3}\right)\left(\dfrac{x_1 + x_2 + x_3}{x_1 + x_2 + x_3 + x_4}\right) + e_2\left(\dfrac{x_1 + x_2 + x_3}{x_1 + x_2 + x_3 + x_4}\right) + e_3$

$e_s = e_1(x_1 + x_2) + e_2(x_1 + x_2 + x_3) + e_3(x_1 + x_2 + x_3 + x_4)$
$|e_s| \leqslant (3x_1 + 3x_2 + 2x_3 + x_4)2^{-t}$

The smallest numbers should be added first.

$|e_s| \leqslant (2x_1 + 2x_2 + 2x_3 + 2x_4)2^{-t}$

Chapter 2

1 The first approximation is the correct solution. If iteration is continued the process will probably diverge due to round-off error.

2 0·567 14 $x_{n+1} = e^{-x_n}$

3 3·692 58

4 0·648 253

5 This is a double root where $f'(x) = 0$ at the root. This introduces large errors in the quotient of the Newton formula.
The method of false position cannot be applied since the function does not cross the axis.

Chapter 3

1 67

2 The root lies between $+2$ and $+3$.

3 $(x^2 - 1·5x + 2)(2x^2 + 10x + 16)$

Chapter 4

1

0·609 375	0·218 75	−0·343 75
0·218 75	−0·562 5	0·312 5
−0·343 75	0·312 5	−0·062 5

2 $-0·428 571$, $0·071 428 5$, $0·428 571$

3 0·47, 0·056, $-0·15$

4
$$L = \begin{bmatrix} 3 & 0 & 0 \\ -6 & 5 & 0 \\ 1 & 8 & 4 \end{bmatrix}$$

5
$$\begin{bmatrix} 1 & 0 & 0 \\ 4 & 1 & 0 \\ 10 & -6 & 3 \end{bmatrix}\begin{bmatrix} 1 & 7 & 5 \\ 0 & 1 & 4 \\ 0 & 0 & 1 \end{bmatrix}$$

6 2·285 714, 0·107 143, 2·142 857

Chapter 5

1 The method is strongly stable. The leading error term is of order h^4.

2 (a) A predictor–corrector method to minimize function evaluations. (b) The Runge–Kutta–Merson. (c) Gear's method for stiff equations.

3 $1\cdot6667$, $2\cdot17778$, $4\cdot6296$, $7\cdot7160$

The problem is started by using the Taylor Series method for the first step followed by the Runge–Kutta–Merson to find the next two starting values. This gives the starting values.

4

x	Prediction	First correction	Second correction
0·2	−0·980 067	starting value only	
0·4	−0·921 076	starting value only	
0·6	−0·825 359	starting value only	
0·8	−0·696 770	−0·696 734	−0·696 731
1·0	−0·540 394	−0·540 328	−0·540 325
1·2	−0·362 462	−0·362 379	−0·362 376
1·4	−0·170 079	−0·169 980	−0·169 979
1·6	0·029 085	0·029 196	0·029 196
1·8	0·227 089	0·227 209	0·227 207
2·0	0·416 040	0·416 163	0·416 159

5

x	Prediction	First correction	Richardson extrapolation
0·2	−0·980 067	starting value only	
0·4	−0·921 076	starting value only	
0·6	−0·825 359	starting value only	
0·8	−0·696 770	−0·696 734	−0·696 736
1·0	−0·540 401	−0·540 335	−0·540 339
1·2	−0·362 478	−0·362 394	−0·362 400
1·4	−0·170 105	−0·170 006	−0·170 013
1·6	0·029 051	0·029 162	0·029 154
1·8	0·227 049	0·227 168	0·227 160
2·0	0·415 996	0·416 119	0·416 111

6 $\alpha_0 = 1, \alpha_1 = 0, \beta_1 = 2$. The leading error term is $-\dfrac{h^3}{3}f'''(x_{n+1})$. The formula is weakly stable.

Chapter 6

1 The entry for $x = 0\cdot6$ should be $9\cdot256$

2 $\mu^2 \equiv (\delta^2 + 4)/4$

3 $y_{n+1} - y_n = \dfrac{h}{12}[5y'_{n+1} + 8y'_n - y'_{n-1}]$

4 $y_n = \frac{5}{4} + \frac{3}{4}(-1)^n - 3(-\frac{1}{2})^n$

5 $y_{n+1} = y_n + h[y'_n + \frac{1}{2}\nabla y'_n + \frac{5}{12}\nabla^2 y'_n + \frac{3}{8}\nabla^3 y'_n \cdots]$

$y_{n+1} = y_n + h[\frac{3}{2}y'_n - \frac{1}{2}y'_{n-1}]$

$y_{n+1} = y_n + \dfrac{h}{12}[23y'_n - 16y'_{n-1} + 5y'_{n-2}]$

$y_{n+1} = y_n + \dfrac{h}{24}[55y'_n - 59y'_{n-1} + 37y'_{n-2} - 9y'_{n-3}]$

Chapter 7

1 3·3043, 6·7042, 8·1819

2 0·5104, 0·2252

3 $2x^2 - 1$, $4x^3 - 3x$, $8x^4 - 8x^2 + 1$
$1 = T_0(x)$, $x = T_1(x)$, $x^2 = [T_2(x) + T_0(x)]/2$
$x^3 = [T_3(x) + 3T_1(x)]/4$, $x^4 = [T_4(x) + 4T_2(x) + 3T_0(x)]/8$

4 $-\frac{11}{32}T_0(x) + \frac{5}{4}T_1(x) - \frac{3}{4}T_2(x)$

5 $6 - 18x + 9x^2 - x^3$, $x^4 - 16x^3 + 72x^2 - 96x + 24$
$120 - 600x + 600x^2 - 200x^3 + 25x^4 + x^5$

Chapter 8

1 $a_0 = -3/5$, $a_1 = 18/5$, $a_2 = 18/5$, $a_3 = -3/5$

2 $h/3$, $4h/3$, $h/3$. Yes.

3 $a_0 = 2$; $a_1 = 2$, $x_1 = -\sqrt{2}$, $x_2 = \sqrt{2}$.

4 0·299 24, 0·303 34, 0·302 28
Richardson extrapolation 0·303 29. Correct value 0·303 29

Chapter 9

1 $\begin{bmatrix} 0·832\,050 & 0 & -0·554\,700 \\ 0 & 1 & 0 \\ 0·554\,700 & 0 & 0·832\,050 \end{bmatrix}$

Eigenvalues 11, 5, −2
Eigenvectors $[6, a, 4]$, $[0, 1, 0]$, $[-2, a, 3]$, a is an arbitrary constant

2 Eigenvalue 48·248 Eigenvector [0·2371, 0·4550, 1·0000]

3 Eigenvalue 5 Eigenvector $[2, 4, -17]$

4 Eigenvalues $+2i$, $-2i$, Eigenvectors $[-\sqrt{2} - 2i, \sqrt{2} - 2i, 2]$, $[-\sqrt{2} + 2i, \sqrt{2} - 2i, 2]$

5 Eigenvalues $2 - \sqrt{2}$, 2, $2 + \sqrt{2}$
Eigenvectors $[1, \sqrt{2}, 1]$, $[-1, 0, 1]$, $[1, -\sqrt{2}, 1]$

6 Eigenvalues 0·038 016, 0·453 835, 2·203 446, 26·304 703
The Eigenvectors are the columns of the following matrix

0·308 686	0·787 275	0·530 366	0·060 187
−0·723 090	−0·163 234	0·640 332	0·201 173
0·594 551	0·532 107	0·391 832	0·458 082
−0·168 412	0·265 358	−0·393 897	0·863 752

Index

Adams–Bashforth formulae, 84, 88, 101, 106, 117
Aitken interpolation, 124–125
Aitken's Δ^2 process, 13, 158

Back-substitution, 60, 64, 66, 74, 78
Backward differences, 111, 113, 117, 118
Bairstow's method, 43, 45–46, 51, 53, 56
Bareiss, E. H., 44, 51
Binary arithmetic, 3–5, 16
Bisection, method of, 27–28, 35, 171

Central differences, 111, 112
Chebyshev approximation, 134
Chebyshev economization, 135–136, 143, 144
Chebyshev oscillation property, 134, 138
Chebyshev polynomials, 119, 130, 131, 132–138, 144
  properties of, 133
Chebyshev quadrature formulae, 153, 154
Chebyshev recurrence relation, 134
Chebyshev series expansion, 136–138, 141
Choleski method, 64, 65
Closed formulae for integration, 84, 85, 101, 102 (*see also* Implicit methods)
Composite quadrature formulae, 150
Consistency, 85
Convergence,
  acceleration of, 12–14
  of iterative methods for linear simultaneous equations, 69–70
  of iterative methods for nonlinear equations, 26–27, 29–30, 32–33
  of Newton–Cotes methods, 148–150
  of predictor–corrector methods, 89–90
Corrector formulae (*see* Predictor–corrector methods)
Cramer's rule, 61

Dahlquist, G., 95, 96, 99
Definite integration, 145–159
  composite formulae for, 150–151
  gaussian formulae for, 151–154, 157, 158

on infinite integrals, 153–154
  Newton–Cotes formulae for, 146–149
  Romberg method of, 154–155, 158
Deflation, of a polynomial, 43
  of a matrix, 165–166, 173
Δ^2 process, 13, 158
Determinant, 57, 61, 160, 161
Diagonal dominance, 70
Difference equations, 86–87, 115–116, 117
Difference table, 108, 109
  errors in, 109, 110, 116–117
Differences, 107–118
  backward, 111, 113, 117, 118
  central, 111, 112
  divided, 122–124, 139
  forward, 107–115, 117
Differential equations, 80–106 (*see* Predictor–corrector methods *and* Runge–Kutta methods)
Differentiation, numerical, 114

Economization of power series, 135–136, 143, 144
Eigenvalues, calculation of, 161–182
  iterative methods for, 162–167, 173–174
  transformation methods for, 167–170
  of tridiagonal matrix, 170, 180
Error, absolute, 7
  in addition and subtraction, 7, 17, 23
  in curve fitting, 119
  in division, 8
  in finite difference tables, 109, 110, 116–117
  in integration formulae, 148–150, 154
  in multiplication, 8
  in predictor–corrector formulae, 85–86, 91
  in solution of linear simultaneous equations, 61–63
Euler's method, 82–83, 89, 103–104
Explicit integration methods, 85, 101–102

False position, method of, 32
Finite differences (*see* Differences)

Floating point arithmetic, 6
Forward differences, 107–115, 117
Fourier series, 120

Gaussian elimination, 58–60, 71, 73, 74
Gaussian quadrature, 151–154, 157, 158, 159
Gauss–Seidel method, 67–68, 71, 75, 78
   convergence of, 69–70
Gear, C. W., 89, 98, 99, 100–101
Givens' method, 168–169, 179–180
Graeffe's method, 49–50

Hermite polynomials, 131
   properties of, 132
Hermite–Gauss quadrature, 154
Hessenberg matrix, 169, 172
Hilbert matrix, 74, 127, 142–143
Householder's method, 169–170

Ill-conditioned linear equations, 58, 61–63,
   126–127
Implicit integration methods, 85, 101–102
Integration (see Definite integration,
      Predictor–corrector methods, and
      Runge–Kutta methods)
Interpolation, 119–125
   Aitken, 124–125
   divided difference, 122–124, 125, 139
   finite difference, 112–114
   iterative, 124–125
   Lagrange, 120–122, 144
   Neville, 124–125, 139
Iterative methods, for a single equation,
      11–12, 25–37
   for an eigenvalue problem, 161, 162–167,
      173–174
   for linear simultaneous equations, 67–70,
      71
      (see also Predictor–corrector methods)

Jacobi iterative method for linear
      simultaneous equations, 67, 71, 74–75
Jacobi polynomials, 133
   properties of, 133
Jacobi transformation method for eigenvalue
      problem, 167–168, 176–179
Jordan elimination scheme, 64, 71

Lagrange interpolation, 120–122, 144
Laguerre polynomials, 131, 144
   properties of, 133
Laguerre–Gauss quadrature, 154, 158
Least squares method, 120, 125–127, 140
Legendre polynomials, 128, 129, 131,
      152–153
   properties of, 133
Legendre–Gauss quadrature, 152–153, 157

Lehmer–Schur method, 48–49, 51
Linear simultaneous equations, 57–79
   direct methods for, 58–66
   iterative methods for, 67–70, 71
   of tridiagonal form, 65–66
L–R method, 171, 180–182

Matrices, deflation of, 165–166, 173
   Hessenberg, 169
   inverse of, 60–61
   lower triangular, 64, 68, 76–77, 171,
      180–182
   orthogonal, 161, 167, 172
   similar, 161, 162, 167
   sparse, 58, 70–71
   symmetric, 161, 167, 168
   triangular decomposition of, 64, 68,
      76–77, 171, 180–182
   tridiagonal, 65, 71, 169, 170–171
Merson (see Runge–Kutta–Merson)
Minimax approximation, 119, 134
Multistep integration methods (see
      Predictor–corrector methods)

Nested multiplication, 39
Neville interpolation, 124–125, 139
Newton–Cotes formulae, 101 (see also
      Definite integration)
Newton's divided difference formula,
      122–124, 139
Newton's forward difference formula,
      113–114, 115
Newton's iterative method, 20–21, 28–31,
      34–35, 44, 46, 51
Nordsieck, A., 88, 89, 100–101
Normal equations, 126
Normalization, 62, 162, 163, 165, 166
Numerical integration (see Definite
      integration, Predictor–corrector
      methods, and Runge–Kutta methods)

Open formulae for integration, 85, 101–102
      (see also Explicit integration methods)
Orthogonal polynomials, 127, 128–132,
      152–154
Over-relaxation, 12–13, 20–21, 68–69, 74, 76

Partial instability (see Stability)
Partial pivoting, 59, 61–62, 71, 121
Polynomials, 107, 109, 119, 120, 126, 160,
      161
   computation with, 32–40
   orthogonal, 127, 128–132, 152–154
   roots of, 39–56
Power method, 162–167, 173–174
Predictor–corrector methods, 82–91, 95–97,
      100–101, 103–104

Process graph, 9

Q–R method, 172
Quadratic factor, 40, 42, 43, 45–46, 53–54, 55
Quadrature (*see* Definite integration)
Quotient–difference algorithm, 47–48, 54–55

Recurrence relation, 129
  for Chebyshev polynomials, 134
Relative error, 7–9
Relaxation factor (*see* Over-relaxation)
Residuals, 63
Richardson extrapolation, 14, 19–20, 90–91, 154–155
Romberg integration, 154–155, 158
Roots of equations, 24–56
Runge, C., 119
Runge–Kutta methods, 89, 91–93, 97,
  compared with predictor–corrector
    methods, 100–101
  stability of, 97–98
Runge–Kutta–Merson method, 89, 92

Scaling, 62
Secant method, 31–32
Simpson's formula, 88, 96, 101, 104, 115, 147, 150, 156–157, 159

Stability, 94–98
  inherent, 94
  partial, 93, 97–98, 104
  of predictor–corrector formulae, 86–88, 95–96
  relative, 94
  of Runge–Kutta formulae, 93, 97–98
  weak, 88, 95–96, 104
  (*see also* Strong instability)
Stiff equations, 97–98, 99–101
Strong instability, 86–88, 95–96, 104
Sturm sequence, 42–43, 52, 170–171, 180, 182
Synthetic division, 39–40, 51–53

Taylor series, 10, 28, 81, 85–86, 88, 99, 103, 104, 135, 136, 143
Thomas algorithm, 66, 71, 77
Transformation methods for eigenvalue
  problems, 167–170
Trapezoidal rule, 83, 89, 101, 103, 115, 147, 149, 150, 154, 159
Triangular decomposition, 64, 68, 76–77, 171, 180–182
Tridiagonal matrices, 65, 71, 169, 170–171

Weak stability, 88, 95–96, 104
Weierstrass, 120